GLOSSARY
OF THE
BRITISH FLORA

GLOSSARY
OF THE
BRITISH FLORA

BY

H. GILBERT-CARTER
M.A. (CANTAB.), M.B., CH.B. (EDIN.)

*Formerly Director of the Cambridge University Botanic Garden
and University Lecturer in Botany*
Honorary Associate of the Linnean Society, London

WITH A PREFACE BY
CHARLES E. RAVEN
D.D., F.B.A.

THIRD EDITION

CAMBRIDGE
AT THE UNIVERSITY PRESS
1964

CAMBRIDGE UNIVERSITY PRESS
Cambridge, New York, Melbourne, Madrid, Cape Town, Singapore,
São Paulo, Delhi

Cambridge University Press
The Edinburgh Building, Cambridge CB2 8RU, UK

Published in the United States of America by
Cambridge University Press, New York

www.cambridge.org
Information on this title: www.cambridge.org/9780521114882

First edition 1950
Second edition 1955
Third edition 1964
This digitally printed version 2009

A catalogue record for this publication is available from the British Library

ISBN 978-0-521-05081-4 hardback
ISBN 978-0-521-11488-2 paperback

IT IS WITH GREAT PLEASURE THAT
I DEDICATE THE THIRD EDITION OF
THE *GLOSSARY* TO A GROUP
OF GENEROUS FRIENDS WHO MADE
ITS PUBLICATION POSSIBLE

NOTE TO THE PREFACE TO THE THIRD EDITION

An opportunity to make available an enlarged new edition of the Glossary has been presented by an occasion which this volume most appropriately commemorates—the eightieth birthday of its author. Some of his friends felt that there could be no better or more fitting tribute to Humphrey Gilbert-Carter's long and continuing interest in the etymology of botanical names than a new edition of his excellent little volume; to enlarge it so that it might be of maximum use to a new generation of students now familiar with 'C. T. & W.' in its second edition seemed to be a valuable and symbolic act. We are pleased to record that the Press, the author, and a group of his friends who have helped to make it possible all find in this a happy and appropriate occasion for celebration.

CONTENTS

PREFACE TO THE THIRD EDITION

By adding about 150 entries to the new edition and improving the definitions of some of the old ones I have attempted to make this edition a fit companion to the second edition of the Clapham, Tutin, Warburg *Flora of the British Isles*.

I have now come to realize fully that most botanists of all countries will continue to pronounce the scientific names of plants as native words. For Italians this is identical, and for Spaniards nearly identical, with Church Latin (p. xii). For southern English it differs widely from all other pronunciations and is often ambiguous or misleading. It is called by Myles the 'Old Method' (p. xx). Recently at a meeting of a learned society a discussion about the dialect names of the woodlouse was in progress. Someone asked what the Latin name was. In the absence of any answer and in spite of my dislike of speaking at meetings, I said that the Latin name was *porcellio*, pronouncing the *c* as *k*. Had I used the 'Old Method', making the *c* an *s*, many of those present would have thought that the word was *porsellio*. Many such misunderstandings may occur. In southern English *caudatus* ('tailed') and *cordatus* ('heart-shaped') are pronounced alike.

Of English dialect pronunciations the less said the sooner mended.

H. G.-C.

HOLCOMBE, DEVON
February, 1964

PREFACE

It is not often that the contributor of 'what men call a preface and prigs a foreword' has as easy a task as this. Of Mr Gilbert-Carter's book two truisms can be truthfully stated. 'It fills a long-felt want' and 'No one else could have done it so well.'

In these days when attention is again being directed to problems of nomenclature and taxonomy and when knowledge of the Latin tongue in spite of demands for it in Responsions and Littlego grows annually less, a handy and accurate guide to the correct pronunciation of the names of our plants has become indispensable. Most botanists know that there is an international code of rules and presumably a committee to keep it up to date: but such knowledge is no help when we are faced with having to pronounce *Menziesia* or *Populus serotina*. Even in the old days when Latin was the language of all educated men it was notorious that the Englishman so mis-mouthed it that no other nation could understand him; and to-day though our 'new' pronunciation has brought us nearer the Continent there are far too many botanists whose naming of species unless written down is quite unintelligible. Mr Gilbert-Carter has not only told us exactly how we ought to speak but has set out the quantity of every syllable. Is there anyone who will read his lists without being constantly convicted of error?

But the book is much more than a guide to pronunciation. It is a full and learned record of the origin and significance of our plant-names—whether they have come down from the ancients, Theophrastus or Dioscorides or Pliny, or commemorate more recent worthies, or are due to some peculiarity of the species, or have been invented as pleasant combinations of letters. That the number of names of which 'origin unknown' is the only comment is so astonishingly small is due to Mr Gilbert-Carter's own very wide range of linguistic knowledge and to the care and zeal with which he has pursued elusive clues and searched the relevant literature. The result has made his book a valuable contribution to the history of botany and a work of real importance to students of philology.

That his long service as a teacher should be thus crowned and that his pupils should receive on his retirement this rich gift and souvenir, will be a great delight and, to some extent at least, a consolation. As one of the multitude of friends who have benefited continually from his help and comradeship I am proud to have been invited to voice what they all will feel, and to say 'Thank you' and 'Well done'.

C. E. RAVEN

Christmas 1949

INTRODUCTION TO THE
FIRST EDITION

This book aims at explaining the meaning, accentuation, and derivation of the generic, trivial, and varietal names of plants mentioned in current British Floras and in the new *British Flora* by Clapham, Tutin and Warburg.

For the purpose of showing the accentuation I have marked the long vowels and left the short ones unmarked. As terminal *o* is always long, except in a few words that do not concern us, I have left it unmarked. It should be noted that all plant-names are treated as Latin words and accented as such. In Latin the accented syllable is stressed. This stress doubtless resembled the accent in English. An example of such stress is furnished by the words *hammer*, with stress on the first syllable, and *begin*, with stress on the second. In a Latin word of two syllables the first syllable is accented. In words of more than two syllables the penultimate syllable is accented if it is long. If the penultimate syllable is short, then the ante-penultimate syllable bears the accent. A 'long' syllable is one containing a long vowel or diphthong, or a vowel followed by two consonants.*

In pure Latin words it may be assumed that a vowel is short if it is followed immediately by another vowel, e.g. *lutĕus, purpurĕus*; but in words transcribed from Greek this rule does not hold. In such words as ACHILLĒA, CENTAURĒA, and HERACLĒUM, the *ē* is contracted from the Greek diphthong ει. A diphthong is treated as a long vowel wherever it occurs, and even when transcribed by a single letter.

---

* Note that in words which have passed into common English usage, the accent is often thrown back. Thus *RESĒDA*, as a generic name, keeps the long, accented penultimate. The 'reseda' of milliners has the penultimate short and the accent on the first syllable. The same applies to *ALYSSUM*, as a generic name, and the 'sweet alyssum' of gardeners.

The names of groups which are called after British genera I have not considered it necessary to include. Groups are usually called after genera by affixing to the stem of the generic name a feminine adjective termination, agreeing with *plantae*, understood. These endings are:

tribe	*-eae*,
subfamily	*-oīdeae*,
family	*-āceae*,
suborder	*-ineae*,
order	*-āles*.

The names of species are binary combinations consisting of a generic name, e.g. SALIX, and a specific epithet, which was formerly called the trivial name, e.g. *alba*. The name so compounded, e.g. *Salix alba*, is known as the specific name.

Since, as already mentioned, all plant names, even when transcribed from Greek, are treated as Latin words, it will not be out of place to give a table of the Greek letters and their Latin equivalents. Since, unfortunately, many students now enter Universities not knowing the Greek alphabet, I have simplified matters by omitting the Greek capitals.

GREEK LETTERS		LATIN EQUIVALENTS	GREEK LETTERS		LATIN EQUIVALENTS
α	Alpha	=a	ν	Nȳ	=n
β	Bēta	=b	ξ	Xī	=x
γ	Gamma	=g	ο	Omīcron	=ŏ
δ	Delta	=d	π	Pī	=p
ε	Epsīlon	=ĕ	ρ	Rhō	=rh, r
ζ	Zēta	=z	σ, s¹	Sigma	=s
η	Ēta	=ē	τ	Tau	=t
θ	Thēta	=th	υ	Hȳpsīlon	=y
ι	Iōta	=i	φ	Phī	=ph
κ	Kappa	=c	χ	Chī	=ch
λ	Lambda	=l	ψ	Psī	=ps
μ	Mȳ	=m	ω	Ōmega	=ō

¹ *s* only at the end of a word.

γ is pronounced and transcribed *n* before κ, γ, χ and ξ.

ζ was probably pronounced *zd*, later *dz*, but is always transcribed *z*, and now usually pronounced *z*.

In this country θ is usually pronounced like *th* in *theology*, ϕ like *ph* in *Philip*, and χ is transcribed *ch* and pronounced *k* (cf. *chemist*). It is better, however, to pronounce χ like *ch* in Scotch *loch*.

The letter υ is the French *u*, German *ü*.

BREATHINGS. The 'rough breathing' (') at the beginning of a word is transcribed and pronounced *h*. The 'soft breathing' (') merely marks the absence of the rough breathing.

DOUBLE CONSONANTS. In syllables which are long because they end in two consonants the vowel is usually pronounced short. Of the Greek letters transcribed by two consonants only ξ (*ks*) and ψ (*ps*) are treated as double consonants. The Greek letter ζ, though transcribed by the Latin *z*, was pronounced *dz* and was formerly treated as two consonants. In transcribed and derived words it has become customary to pronounce this letter as *z*. The English words *rhizome* and *ōzone* were formerly pronounced *rhĭdzome* and *ŏdzone*. The Greek letter ρ, when initial, was always aspirated and written with a rough breathing ($\dot{\rho}$), so that at the beginning of words it is transcribed *rh*. When it is the first letter of the second element of a compound word ρ is doubled.

DIPHTHONGS. A Latin or Greek diphthong is a coalition of two vowel sounds pronounced in one syllable. The common Latin diphthongs are *ae*, *au*, and *oe*. The combination *eu* is a diphthong in some few Latin words, none of which, however, occurs in this Glossary. Many Latin adjectives, for example, end in *eus*, which is pronounced as two syllables. When *eu* is a transcription of the Greek $\epsilon\upsilon$ it is pronounced as a diphthong, i.e. in one syllable. In order to understand which pairs of Greek vowels form diphthongs it is necessary to know that α, ϵ, η, o, and ω are called hard vowels, and that ι and υ are called soft vowels. When a hard vowel precedes a soft vowel a diphthong is commonly formed, as for example $\epsilon\upsilon$, mentioned above.* When a soft vowel precedes a hard vowel no diphthong is formed, so that the word *sophia* ($\sigma o\phi\acute{\iota}\alpha$) has three syllables.

The Greek diphthong $\alpha\iota$ is transcribed *ae*, $\epsilon\upsilon$ is transcribed *eu*, $o\iota$ is transcribed *oe*, and $o\upsilon$ is transcribed *u*, which is always long, because it is a contraction of a diphthong, and $\epsilon\iota$ is transcribed *e* or *i*, which are long for the same reason.

GREEK ACCENTS. Nearly all Greek words are written with an accent, ` ´ ^, which originally marked the raising of the tone of voice and not stress or tonic accent like that of Latin or English. Students of Botany may well disregard the Greek accents, though they may be of importance in distinguishing words of different meanings but similarly spelt, as $\theta\epsilon\rho\mu\acute{o}s$, *hot*, and $\theta\acute{\epsilon}\rho\mu o s$, *lupin*. It should be noted that the circumflex, ^, occurs only on long vowels and diphthongs.

COMPOUND WORDS. When two elements are joined together to form a word—many plant names are so formed—the second element is added to the *stem consonant* of the first. A good example is DENTARIA. The first element of this word is *dens*, tooth, stem *dent-*. The omission of the stem consonant in such words should be forbidden by rules of nomenclature. The spelling LEONURUS for LEONTURUS is slovenly, and CALYSTEGIA for CALYCOSTEGIA is both slovenly and misleading.

Usually a joining vowel is necessary. In Greek words ŏ is the ordinary joining vowel and in Latin words ĭ; but ŏ is common in Late Latin, and is used in some Latin plant names.

The Greek endings *-os* and *-ov* commonly become Latin *-us* and *-um*. Words ending in *-η* nearly always appear in Latin with *-a*.

After reading these notes students should find themselves able to accentuate any of the names in the Glossary. For actual pronunciation of the letters they would do well to use the restored pronunciation of Latin as taught in most schools. I make this recommendation because Latin is an international language,

* In the suffix *-oīdēs* the *o* and the *i* are in separate syllables. See p. 58.

which, in the restored pronunciation, is understood, up to a point, by all educated people wherever European civilization has spread. But to those who wish to pronounce plant names in other ways I would say: 'Please do not be angry with me, I shall love you just as much however you pronounce the names.' Personally I vary my pronunciation to suit my audience and my feelings. For the anglicized pronunciation of plant names rules exist, but they are often disregarded or yield to usage.

Since this book is the first of its kind, and since its author is not competent for the task of writing it, imperfections must abound. I look forward to the time when someone better qualified than myself may write a bigger and better book. I hope that to those who wish to advance this kind of learning my Bibliography will be useful. There are many other books to be consulted, but those I have mentioned are the ones I have found most useful. All books on this subject should be consulted with caution.

Of the scholarly footnotes in the Floras of Ascherson-Graebner and Hegi I cannot speak too highly. I have made free use of them. All students of plant names should be acquainted with Mrs Arber's *Herbals*.

I should like to end with two recommendations, which, if accepted, might help to standardize the pronunciation of plant names.

In this country the termination -*on* is usually pronounced short whether it represents the Greek masculine termination -ων or the neuter -ον. This treatment is not to be recommended.

In the Latin termination -*inus*, -*a*, -*um* the *i* is usually long. It is short in *annotinus*, *serotinus*, and a few other words. In the Greek termination -ινος, -η, -ον the ι is short. I recommend that if the name ending in -*inus* is a genuine Greek word (as *amygdalinus*), the *i* be kept short. If it is a fabricated word, as *calycinus*, it should be treated as a Latin word with a long *i*.

My thanks are due to many helpers. The chief of these is Canon Raven. One day, when he was in the turmoil of his duties as Vice-Chancellor, I mentioned to him that I had made the card index that forms the basis of this Glossary. He very kindly offered to read it through, and when he had done so told me that he thought it was worth publishing. My feelings were those of Dr Johnson, who, when asked whether he had replied to a compliment paid him by the king, answered: 'No sir. When the king had said it, it was to be so. It was not for me to bandy civilities with my sovereign.' I am glad to be able to record this splendid work of supererogation performed by a Vice-Chancellor. My experience of overworked University Officers leads me to believe that the more overworked they are the more willingly do they lend a helping hand to those in need of their assistance. I am deeply grateful to Canon Raven for reading the proofs of the Glossary. He, without rival, because of his combined knowledge of the classical languages with their herbalistic transmogrifications, and of his wide and detailed acquaintance with the British Flora, is the Agamemnon (ἄναξ ἀνδρῶν) for this task. I am glad that the proof reading and the writing of the kind Preface fell on his shoulders during a period of leisure. Our Public Orator, Mr W. K. C. Guthrie, whose Latin speeches are understood by people of all nations and tongues, has given me valuable guidance on matters dealt with in the Introduction. Mr H. M. Adams, Librarian of Trinity College, Cambridge, has kindly guided my faltering feet along several difficult paths. I am indebted for much help to Mr H. S. Marshall, Librarian to the Kew Herbarium, and to other members of the Herbarium staff, a group of busy experts, always willing to help botanists through their difficulties. Mr S. Max Walters of St John's College, Curator of the University Herbarium, has given me valuable hints. While the book was in the Press, Professor N. B. Jopson, of St John's College, kindly helped me with several passages of the Introduction that were still causing me difficulty. I am also grateful to Professor T. G. Tutin for last-minute additions and corrections.

H. G.-C.

INTRODUCTION TO THE
SECOND EDITION

The first task in preparing the second edition of my *Glossary* was to correct numerous errors, which, through my carelessness or ignorance, were contained in the first edition. Had it not been for the kindness of several learned friends, many of these faults would have remained unemended.

Among these kind friends I am chiefly indebted to Dr T. A. Sprague. Those familiar with Dr Sprague's knowledge of the various branches of plant nomenclature will be less surprised at his insight into my defections, than that I—and not he—had the temerity to write the Glossary. He has very kindly read the proofs of this edition.

While emphasizing the debt that this edition owes to Dr Sprague, I should like to explain that he is not responsible for the spelling of the plant names. Dr Sprague strictly adheres to the rules of spelling proposed in the International Rules of Plant Nomenclature. My spellings do not always agree with these Rules; but they will be found to agree closely with those of Hegi's *Flora von Mittel-Europa*, the standard flora of Central Europe for many years to come.

I am also deeply grateful to various members of the staff of the Kew Herbarium, especially to Mr Noel Sandwith and Mr H. K. Airy Shaw, to Mr W. T. Stearn, Librarian to the Royal Horticultural Society, and to Mr David McClintock.

The second task was to compare the Glossary with the pages of the new Flora, published in 1952. This resulted in the addition of about 200 entries, bringing the total up to over 2000. These additions will, I hope, make the Glossary a worthy handmaiden to the great Flora. Mr Noel Sandwith was the chief among those who helped me to elucidate the more obscure of these words.

I have omitted many purely botanical terms such as *cymosus* and *racemosus*. Those unfamiliar with such epithets should refer to 'cyme' and 'raceme' in the excellent glossary at the end of the Flora. Of personal and geographical names I have given only a selection. It is needless to state that *europaeus* means European and that *americanus* means American.

While adding names in order to bring the Glossary closer to the new Flora, I have retained, and even added, certain words found only in older books. Such a misunderstood word as *Serrafalcus*, still in frequent use among agrostologists, deserves to be properly glossed.

How much this book owes to the Cambridge University Press it is needless to tell. It was because of their work, and not mine, that the first edition was one of the books selected by the National Book League for their Exhibition of British Book Design held in March 1951. Many of the corrections and improvements in this edition were prompted by the suggestions of a keen and sedulous press reader. It was at the instance of this reader that I altered the spelling of the names of the Greek letters in the table on p. xii, making them all transcriptions of their Greek names. This alteration made these names consistent with the rules of transcription already stated. Some of the Greek letters are used as mathematical symbols, and have, in this country, their own English pronunciation. Let it not be thought that I should wish English-speaking scientists to pronounce the names of these letters otherwise than they have always done. We all bear in mind the Vicar of Crewe,

> Who kept a tom-cat in a pew;
> He taught it to speak
> Alphabetical Greek,
> But it never got further than μ.

The names of those who have helped me with individual words are too numerous to record. Fools proverbially attempt tasks which angels hesitate to undertake. The reason for this may well be that when a fool attempts a task beyond his powers, he finds that he can rely upon limitless angelic assistance.

There are several difficulties in the pronunciation of certain plant names that I cannot solve. One of these is how to pronounce words derived from personal names, belonging to languages with sounds differing widely from those of Latin; HUTCHINSIA and SCHKUHRIA are examples.

Another kind of difficulty is presented by such names as LISTERA and LAVATERA, in which the sounds can be latinized, but the quantity of the penultimate vowel is doubtful. For the sake of euphony I have chosen to lengthen this vowel.

I am unable to follow the authors of the Flora in their wholesale decapitalization of specific epithets, especially those specific epithets which are, or were, generic names. Such combinations as *Campanula medium* and *Sedum rosea* will appear to those unacquainted with botanical Latin as glaring false concords. Further, my old-fashioned respect for my elders and betters is shocked by such discourteous frivolities as *smithii* and *brownii*.

A kind reviewer of my first edition, to whom I am exceedingly grateful, called attention to an important omission from my Bibliography, viz. the 'Pronouncing Dictionary' by Myles, in Nicholson's *Dictionary of Gardening*. This article, 99 pages in length, was published in 1889. It is of great interest, and, as my reviewer states, 'seems to have been the first attempt in the English language seriously to examine the subject'. Because of the historical interest of Myles's 'Pronouncing Dictionary', I append the following tables, in which the author contrasted what he called the Old Method, which, as he said, was then happily becoming obsolete, with the Accurate Method, which was taking its place. Myles's Correct Method may be useful to those unacquainted with the French and German sounds to which I have referred in my table on p. xxiii. His Old Method gives approximate rules for the anglicized pronunciation of plant names.

OLD METHOD

Vowels

a, short, as in făt.

e, „ „ slĕnder.

i, „ „ thĭn.

o, „ „ rŏtten.

u, „ „ stŭbborn.

y, „ „ cўnical.

a, long, as in bāther.

e, „ „ ēvil.

i, „ „ īce.

o, „ „ vōter.

u, „ „ mūle.

y, „ „ cўpher.

Diphthongs

æ
œ } as *ee* in feed.

ei as in the word *eye*.

au as *aw* in bawl.

Consonants

c and g hard before a, as in cats, gaping.

c and g „ „ o, „ cows, goring.

c and g „ „ u, „ cud, gulping.

c and g soft before e, „ central, gentleman.

c and g „ „ i, „ circular, gin.

c and g „ „ y, „ cynical, gymnast.

CORRECT METHOD

Vowels

a, short, as in ăpart.

e, „ „ slĕnder.

i, „ „ thĭn.

o, „ „ rŏtten.

u, „ „ powerfŭl.

a, long, as in psālmist.

e, „ „ vēined.

i, „ „ machīnist.

o, „ „ vōter.

u, „ „ rūler.

y is sounded nearly like i, but the exact sound cannot be rendered in English.

Diphthongs

æ
œ } nearly as *ai* in pain.[1]
ei

au as *ou* in house.

Consonants

g always hard, as in good.

c „ „ „ muscular.

ch „ „ „ Christian.

---

[1] It is difficult to understand how Myles reconciled his pronunciation of the first two of these diphthongs with his Correct Method of pronouncing the Latin vowels. Latin *a* and *e* pronounced in one syllable sound nearly as *i* in English *mine*. Similarly *o* and *e* sound nearly as *oi* in oil.

The pronunciation of so-called 'Church Latin' differs from the 'restored' pronunciation (p. xxiv) chiefly in the sounds of *c*, *g*, and *v*. In Church Latin, *c*, when standing before *e*, *i*, or the diphthong, *æ* or *œ*, has the sound of *ch* in English 'child'. Similarly *g* before these letters is pronounced as *g* in English 'giant'; *v* is pronounced as *v* in English 'very'.

We have then at least three methods of pronouncing Latin: the Restored Method (pp. xx, CORRECT METHOD, and xxiv); the Anglicized Method (pp. xx, OLD METHOD); and Church Latin. All three methods may be regarded as 'right' provided the laws of Latin accentuation (p. xi) are obeyed.

H. G.-C.

April 1955

APPROXIMATE 'RESTORED' PRONUNCIATION OF THE LATIN LETTERS

ā   as *a* in English *father* or German *Vater*.

ă   the same sound pronounced short.

b   as in English.

c   always as English *k*, but without aspiration.

d   the English *d* will suffice.

ē   as in French *été* or in German *See*. Avoid the diphthong in English *say*.

ĕ   approximately the English *e* in *fret*.

f   as in English.

g   always as in English *get*.

h   as in English.

ī   nearly as *ee* in English *seed*.

ĭ   *i* in English *fit* will do.

j   (consonantal *i*) as *y* in English *yet*.

l   as French or German *l*. Avoid the *l* in English *wall*.

m   as in English.

n   the English *n* will do.

ō   German long *ō* is usual. Avoid the diphthong in English *stone*.

ŏ   as in English *got*, or better, German *Gott*.

p   as in English, but without aspiration.

q   always followed by *u* and pronounced as *qu* in English *quite*.

r   always trilled as in Scotland.

s   as in English *cats*; never voiced as in *dogs*.

t   French *t*. The English sound is aspirated, but will do.

ū   as English *oo* in *shoot*.

ŭ   as English *u* in *full*.

v   consonantal *u*; *w* in English *week* will do.

x   as in English.

For letters borrowed from Greek see p. xiii.

Double letters should be pronounced double, e.g. *nn* as in *penknife*, not as in *Pennine*, *cc* as in *bookcase*, etc.

GLOSSARY

a-, ἀ-, prefix expressing want or absence. This prefix is commonly called ἀ privative.

Abiēs, Latin name of a coniferous tree.

abrotonoīdēs, resembling *Artemisia Abrotonum*; ἀβρότονον, name, in Greek authors, of ARTEMISIA spp. and *Santolina Chamaecyparissus*.

Absinthium, name of a plant in Cato, etc. (ἀψίνθιον in Xenophon).

Acaena, ἄκαινα (ἀκή, ἀκίς), thorn, from the sharp spines on the calyx of some species.

Acanthium, ἀκάνθιον, name, in Dioscorides, of a thistle-like plant, from ἄκανθα, prickle.

Acanthus, ἄκανθος, name of *Acanthus mollis* (ἄκανθα, thorn, prickly plant).

acaulis, e, also **os, on,** stemless, ἀ privative and *caulis* (καυλός, stem).

Accipitrīna, word formed from *accipiter*, hawk, on analogy of HIERA-CIUM, q.v.

Ācer, name, probably, of *A. campestre*. Note short *a* and neut. gender.

ācer, cris, cre, (sharp, pointed) acrid, i.e. with a sharp, burning, pepper-like taste. A masc. in *-cris* is often used in botanical Latin.

Aceras, ἀ privative and κέρας, (horn) spur, from the spur-less perianth.

acerbus, a, um, harsh to the taste.

Acētosella, word formed from *acētum*, vinegar, referring to the acid taste.

acētōsus, a, um, medieval Latin adj., meaning acid, from *acētum*, vinegar.

Achillēa, ἀχίλλειος, plant-name used by Greek authors. In Pliny *mille-folia* is a synonym of *achillēa* and *achillēos* (named after Achilles, pupil of Chiron).

Achras, ἀχράς, a kind of wild pear, perhaps *Pirus amygdaliformis*, a common Mediterranean species.

Achyrophorus, ἄχυρον, chaff, and -φόρος, bearing, referring to the membranous scales on the receptacle.

aciculāris, e, needle- or pin-like (*acicula*, diminutive of *acus*, needle).

acidus, a, um, acid, sour, i.e. tasting like vinegar or lemon juice and reddening blue litmus (*aceo*, be sour).

acinifolius, a, um, having leaves resembling those of ἄκινος (*Acinos arvensis*).

Acinos, ἄκινος, name of a plant in Dioscorides (*Calamintha graveolens*).

Aconītum, the Latin name of species of this genus (ἀκόνιτον, of doubtful etymology).

Acorus, ἄκορον, name, in Dioscorides, of *Iris Pseudacorus*.

ācris, fem. sometimes used also for masc. of *ācer*.

Acrotonae, ἄκρα, summit, apex, and τόνος, gerund form of τείνω, stretch, extend. The apex of the anther loculus is prolonged against the rostellum.

Actaea, name in Pliny, said to be connected with ἀκταία (ἀκτέα), elder-tree. The leaves of ACTAEA somewhat resemble those of SAMBUCUS.

actino-, having rays (ἀκτίς, ῖνος, ray); the stem vowel is shortened in compounds used as ordinary English words.

acūleātus, a, um, thorny, prickly (*aculeus*, spine, prickle).

adenocaulon, Greek neut.adj. formed from ἀδήν, ένος, gland, and καυλός, stem.

Adiantum, ἀδίαντον, name, in Greek authors, of a plant with unwettable leaves, perhaps *A. Capillus Veneris* (ἀδίαντος, unwetted).

Adōnis, used as a plant-name by Mattioli. The flower which sprang up from the blood of Adonis, favourite of Aphrodite, was called *Adōnium*.

Adoxa, ἀ privative and δόξα, glory, referring to the insignificance of the plant.

adpressus, a, um, appressed, term applied to structures (hairs, pods, etc.) which lie flat against the axis bearing them (*ad*, on, against, and *premo, pressum,* press).

adscītus, a, um, derived, assumed (p.part. of *a(d)scisco*).

Aegilōps, αἰγίλωψ, name of several plants, probably including a wild oat. The first element is doubtless αἴξ, αἰγός, goat. See Hegi, vol. I, p. 390, footnote, and p. 254, under *Avena sativa*.

Aegīrus, a name in Greek authors of *Populus nigra* (αἴγειρος).

Aegopodium, from αἴξ, αἰγός, goat, and πόδιον, little foot, referring to the shape of the leaves.

aemulus, a, um, striving, rivalling.

Aesculus, name of a tree, perhaps *Quercus petraea*.

aestīvālis, e, = *aestivus*.

aestīvus, a, um, of (flowering in) summer (*aestas*, summer).

Aēthionēma, origin much disputed, perhaps from ἀήθης, unusual, and νῆμα, thread, from the winged or toothed filaments.

Aethūsa, name first used by Linnaeus, from αἴθω, burn, referring to the shining foliage. Cf. *Gleiss(e)*, the German name of *Aethusa Cynapium*, from *gleissen*, glitter (αἴθουσα (sc. στοά) meant a veranda to catch the sun).

Agathophytum, ἀγαθός, good, and φυτόν, plant, referring to *Chenopodium bonus-Henricus*.

Agraphis, ἀ privative and γράφω, write. See *nōn-scriptus*.

4

agrārius, a, um, growing in fields, lit. pertaining to land (*ager*, field).

agrestis, e, wild, usually means growing on arable land (*ager*, field).

Agrimōnia, plant-name in texts of Celsus and Pliny. The word, however, is a false reading for *argemonia*, name of a plant similar to *argemōnē*. *Argemon* (ἄργεμον) was the name of a plant called in pure Latin *lappa canāria*.

Agropȳrum, ἀγρός, field, and πῦρος, wheat. Better spelt, as by Hegi, AGRIOPYRUM (ἄγριος, wild).

Agrostemma, ἀγρός, field, and στέμμα, garland, suited for garlands of wild flowers. The word was formed by Linnaeus by analogy with CORONARIA, q.v.

Agrōstis, ἄγρωστις, name of a plant in Theophrastus and other authors, thought to be *Cynodon Dactylon*.

Aira, αῖρα, name of a weed among wheat, thought to be *Lolium temulentum*.

Aïzōāceae, from the genus AïZOON. The word ἀείζωος means everlasting, and τὸ ἀείζωον was the name of an evergreen plant, probably a SEMPERVIVUM (ἀεί, for ever, and ζωή, life).

aïzōīdēs, like AïZOON, a name now used for a genus of *Aïzoaceae*.

Ājācis, gen.sing. of Ajax, from whose blood sprang a flower bearing the initial letters of his name, the Greek capitals AIA.

Ajuga, a name used by Scribonius Largus, physician, as a synonym of *abiga*, a plant, in Pliny, producing abortion (*abigo*, procure abortion).

albescens, becoming white.

albidus, a, um, white.

albus, a, um, white.

Alcea, ἀλκαία (ἀλκέα), name, in Dioscorides, probably of *Malva moschata*.

Alchemilla (Alchimilla), name of our genus in Hieronymus Braunschweig, origin unknown.

Alisma, ἄλισμα, name of a water-plant, used by Dioscorides, probably not belonging to the genus ALISMA.

Alliāria, name used by Fuchs, referring to the garlic-like smell of the plant (*allium, alium*, garlic).

Allium, *alium,* garlic.

Allosōrus, ἄλλος, other, different, and σωρός, heap (sorus), from the variously shaped sori.

Alnus, the Latin name of *Alnus glutinosa*.

aloīdēs, resembling species of the genus ALOË.

Alōpecūrus, ἀλώπηξ, εκος, fox, and οὐρά, tail, from the brush-like inflorescence.

alpester, tris, tre, usually synonymous with *alpīnus*.

alpigena (substantive), something produced in the Alps (*alpes*, high mountains, and *gigno*, bring forth).

alpīnus, a, um, growing in alpine places (*Alpes*, high mountains). N.B. The German word *Alpe* (pl. *Alpen*) means: (1) the Alps; (2) pastureland on mountainside, alpine meadow.

Alsine, ἀλσίνη, name of a plant in Theophrastus, perhaps *Parietaria cretica*.

Althaea, ἀλθαία, name of a plant in Theophrastus (see also appendix).

altilis, e, nutritive, nourishing (*alo*, nourish).

altissimus, a, um, superlative of *altus*.

altus, a, um, high, tall (p.part. of *alo*, nourish).

Alyssum, ἄλυσσος, curing rabies, from ἀ privative and λύσσα, rage, fury, rabies, also a worm under the tongue of dogs said to cause rabies.

Amarantus, name of a plant in Latin authors, ἀμάραντος in Nicander, from ἀ privative and μαραίνω, fade away.

Amārella, fem. diminutive of *amārus*, bitter.

amārus, a, um, bitter, i.e. tasting like quinine or hops.

Amaryllidāceae, called after the genus AMARYLLIS (the name of a country girl in Theocritus and Virgil).

Ambrosia, ἀμβροσία, name in Dioscorides of *Ambrosia maritima*. The word means immortality or the elixir of life.

Amelanchier, the Provençal name, of doubtful origin, of *A. ovalis*.

Āmentiflōrae, fem.pl.adj. (sc. *plantae*) from *āmentum*, thong (catkin), and *flōs, ōris*, flower.

amethystēa, of the colour of the precious stone amethyst.

Ammī, name in Greek and Latin authors of *Trachyspermum (Carum) copticum*.

Ammophila, sand-lover, ἄμμος, sand, and φίλος, friend. Also used as adjective, *ammophilus, a, um*, sand-loving.

Amōmum, ἄμωμον, name of an Indian spice plant, used for our species of SISON because of its powerful aroma.

Ampeloprasum, ἀμπελόπρασον, name in Dioscorides of a species of ALLIUM which probably grew in vineyards, from ἄμπελος, vine, and πράσον, leek.

amphibius, a, um, living a double life, i.e. both on land and in water, from ἀμφί-, both, and βίος, life.

amplexicaulis, e, having the base of the leaf or petiole partially embracing the stem (*amplexus*, encircling, embracing, and *caulis*, stem).

ampullāceus, a, um, like a flask or *ampulla*, as the perigynia of *Carex ampullacea*. See Rich, p. 30.

amygdalinus, a, um, ἀμυγδάλινος, of almonds. The leaves of *Salix triandra* taste and smell of sweet almonds. The ι is short in the Greek word, but it is, perhaps, permissible to lengthen it in the Latin form.

amygdaloīdēs, almond-like.

6

Amygdalus, ἀμύγδαλος, almond-tree.

Anacamptis, from ἀνακάμπτω, bend back, referring to the long, curved spur.

Anagallis, name, in Dioscorides, probably from ἀν-, without, and ἀγάλλομαι, boast, referring to an unpretentious plant.

anagyroīdēs, resembling the genus ANAGYRIS.

Ananassa, name of the pine-apple (*Ananas sativus*), probably of Peruvian origin.

Anaphalis, origin obscure, perhaps a bad anagram of GNAPHALIUM.

Anastatica, fem.adj. from ἀνάστασις, rising up, resurrection. *A. hiero-chuntica* (rose of Jericho) during the dry season withers away and contracts into a ball of wickerwork, which blows about in the breeze. When it reaches a wet place, or when rain falls, it expands.

Anchūsa, ἄγχουσα or ἔγχουσα, name in Aristophanes of a plant, perhaps *Alkanna tinctoria*, yielding a red dye used as rouge.

andegavensis, e, of Angers in Anjou (*Andegava*).

andevagensis, e, mis-spelling of *andegavensis*.

Andromeda, Ἀνδρομέδη, daughter of Cepheus and Cassiope.

Androrchis, ἀνήρ, ἀνδρός, man, and ORCHIS, referring to *Orchis mascula*.

Androsaemum, ἀνδρόσαιμον, name, in Dioscorides, of a plant with blood-red juice, from ἀνήρ, ἀνδρός, man, and αἷμα, blood.

-andrus, a, um, -stamened, e.g. *triandrus*, 3-stamened (ἀνήρ, ἀνδρός, man, i.e. stamen).

Anemōnē, ἀνεμώνη, *Anemone coronaria*, a common Mediterranean species, from ἄνεμος, wind, and -ώνη, feminine patronymic suffix, 'daughter of the wind'. Connexion with wind obscure. Some consider the name to be a Greek loan word of Semitic origin, not connected with ἄνεμος, but introduced through the cult of Adonis (Naaman), from whose blood *Anemone coronaria* was believed to have sprung. In Britain the word is usually pronounced as the English name, with short *o* and accented antepenultimate.

Angelica, name in Matthaeus Sylvaticus of *Archangelica officinalis*, which was said to have been revealed by an angel (*herba angelica*, angelic herb).

angiocarpus, a, um, ἀγγεῖον, vessel, urn, and καρπός, fruit.

Angiospermae, ἀγγεῖον, vessel, and σπέρμα, seed, referring to the closed carpels.

anglicus, a, um, English.

anglōrum, gen. of *Anglī*, the English people. *Hieracium anglorum* is thought to be endemic to England.

angusti- (*angustus, a, um*), narrow-.

Anisantha, ἄνισος, unequal, uneven, and ἄνθος, flower. In the specimen described the lowest flower of the spikelet is fertile, the next one staminate, and the rest sterile and embracing the rachilla.

Anisophyllum, ἄνισος, uneven, and φύλλον, leaf. The leaves of species of this section are oblique at the base.

annōtinus, a, um, of last year. Note the long *o* and short *i*. The annual shoots of *Lycopodium annotinum* are clearly demarcated.

Anōgramma, ἄνω, up, upwards, and γραμμή, line (sorus). The sori first appear at the apices of the young pinnae.

Anoplobatus, ἄνοπλος, unarmed, and *-batus*, q.v., referring to absence of thorns.

anserīnus, a, um, pertaining to geese, hence, growing on goose-greens (*anser*, goose). In Linnaeus's *Flora Suecica*, two Swedish names are given to *Potentilla anserina*: Gåsört (goose-weed), and Silfverört (silver-weed).

Antennāria, from the *antenna*-like pappus hairs (*antenna*, sailyard).

Anthemis, ἀνθεμίς = ἄνθος, flower. The name is used by Dioscorides for a plant called also λευκάνθεμον and χαμαίμηλον.

-anthemus, a, um, -flowered, as *polyanthemus, a, um,* many-flowered (ἀνθεμίς, flower).

-anthēs, -anthus, a, um, -flowered (ἄνθος, flower).

Anthoxanthum, ἄνθος, flower, and ξανθός, yellow, from the colour of the ripe spikelets.

Anthriscus, ἄθρυσκον, name in Greek author of *Scandix australis*.

anthrōpophorus, a, um, man-bearing, ἄνθρωπος, man, and *-φόρος,* bearing (φέρω, bear), from the form of the flowers.

Anthyllis, ἀνθυλλίς, name of a plant in Dioscorides.

Antirrhīnum, ἀντίρρινον, name, in Dioscorides, of *Antirrhinum majus,* from ἀντι-, counterfeiting, and ῥίς, ῥινός, nose. The description in Theophrastus fits *Valantia hispida.*

Apargia, ἀπαργία, name of a plant in Theophrastus, perhaps from ἀπό, denoting origin, and ἀργία, lapse of cultivation (fallow land).

Aparīnē, ἀπαρίνη, name of *Galium Aparine* in Theophrastus and Dioscorides.

Apera, equivocal word invented by Adanson.

apetalus, a, um, without petals, from ἀ privative and πέταλον, leaf, used in botany for petal.

Aphaca, ἀφάκη, name of a leguminous plant (cf. φακῆ, lentil porridge, and φακός, lentil plant and seed).

Aphanēs, ἀφανής, unseen, unnoticed. The species are inconspicuous.

8

aphyllus, a, um, leafless, from ἀ privative and φύλλον, leaf.

apiculātus, a, um, furnished with a short, acute, but not stiff, point (*apiculus*, diminutive of *apex*, point, summit).

apifer, a, um, bee-bearing (*apis*, bee, and *fero*, bear).

Apium, name used by Latin authors for several umbelliferous plants.

Apocynāceae, fem.pl.adj. (sc. *plantae*) formed from APOCYNUM, the type genus of the family. The ἀπόκυνον of Dioscorides was *Marsdenia erecta*, 'dog's-bane', from ἀπο-, asunder, and κύων, κυνός, dog.

Aponogētōn, word formed on analogy with POTAMOGETON, the first element being, apparently, ἄπονος, without toil or trouble.

appropinquātus, a, um (p.part. of *appropinquo*, come near, approach), approaching, in appearance, some other species.

apterus, a, um, wing-less, from ἀ privative and πτερόν, wing.

aquāticus, a, um, growing in water (*aqua*, water).

aquātilis, e = aquāticus.

Aquifolium, name, in Pliny, of *Ilex aquifolium*, from *aquifolius, a, um,* having pointed leaves, from *acus*, needle, and *folium*, leaf.

Aquilegia, perhaps from old forms of the German name *Akelei*. Hegi (III, p. 480) gives twelfth-century spellings *Acheleia* and *Agleia*.

aquilīnus, a, um, to do with eagles. The bundles of the rhizome, on section, resemble a spread eagle (*aquila*, eagle) (see appendix).

Arabidopsis, resembling species of the genus ARABIS (ὄψις, appearance).

Arabis, origin obscure.

arachnītēs, spider-like, from ἀράχνης, spider, and -ίτης, connected with.

Arachus, ἄραχος, name, in Galen, of a plant thought to be *Vicia Sibthorpii*.

Aracium, name formed by Necker on analogy with HIERACIUM, from ἄρακος, a Tyrrhenian word for ἱέραξ, hawk.

Araliāceae, called after the genus ARALIA (origin of name obscure).

arānifer, a, um, spider-bearing, from *arānea*, spider, and *fero*, bear.

araucānus, a, um, of the province Arauco in S. Chile.

Araucāria, adjective from Arauco, name of a province in S. Chile.

arbuscula, a little tree (diminutive of *arbor*, tree).

Arbutus, the Latin name. Note the short first *u*.

Archangelica, said to have been revealed by the Archangel Gabriel.

Archichlamydeae, fem.pl.adj. (sc. *plantae*), from ἀρχή, beginning, and χλαμύς, cloak (perianth), referring to the low grade of elaboration of the perianth.

Arctium, name in Pliny of a plant also called *arcturus* (see appendix).

Arctostaphylos, ἄρκτος, bear, and σταφυλή, bunch of grapes, translation of *ūva ursī*.

Arctōus, ἀρκτῷος, of a bear (ἄρκτος, bear).

9

arcuātus, a, um, bent like a bow (arcus, bow).
Aremōnia, transformation of AGRIMONIA, q.v.
Arēnāria, see arenārius.
arēnārius, a, um, growing on sand ((h)arēna, sand).
Argemōnē, ἀργεμώνη, name, in Dioscorides, of a poppy-like plant.
argenteus, a, um, silver-coloured (argentum, silver).
argyro-, silver- (ἄργυρος).
Aria, ἀρία, name of a plant in Theophrastus, perhaps QUERCUS sp.
aristātus, a, um, furnished with an awn (arista).
Aristolochia, ἀριστόλοχεια (-λοχία in Theophrastus), name of species of
   ARISTOLOCHIA, from ἄριστος, best, and λόχος, birth, i.e. used to
   assist childbirth. The species are poisonous, and may cause abortion.
   Note the 'birth signature' of the perianth. The inflated base repre-
   sents the uterus and the tube the birth passages.
Armeria, from the French names armoires and armoiries applied to
   species of DIANTHUS with aggregated flowers.
Armoracia, ἀρμορακία, name, perhaps, of Raphanus Raphanistrum.
Arnoseris, ἄρνος, lamb, and σέρις, kind of chicory or endive. See -seris.
Arrhenatherum, ἄρρην, ενος, Attic for ἄρσην, male, and ἀθήρ, έρος, awn.
   The spikelets are 2-flowered; the upper flower is hermaphrodite; the
   lower one is male, and its lemma bears a long awn.
arrhīzus, a, um, rootless, from ἀ privative and ῥίζα, root.
Artemisia, ἀρτεμισία, name, in Dioscorides, of a plant, called after
   Artemis (Diana).
arthro-, ἄρθρον, joint.
Arthrocnēmum, ἄρθρον, joint, and κνήμη, leg, shank, internode. The
   shoots resemble those of SALICORNIA.
Arthrolobium, ἄρθρον, joint, and λοβός, pod.
articulātus, a, um, jointed.
Arum, ἄρον, name, in Theophrastus, of Arum italicum.
arundināceus, a, um, reed-like ((h)arundo, inis, reed).
Arundo, reed, probably Arundo Donax (harundo).
arvālis, e, growing on arable land (arvum).
arvensis, e, growing on arable land, a common trivial name for plants
   found on ploughed fields (arvum (solum), arable land).
Asarum, ἄσαρον, name, in Dioscorides, of A. europaeum.
Asparagus, ἀσπάραγος, the name of 'Asparagus' and the shoots of
   similar plants.
asper, era, erum, rough.
aspernātus, a, um, despised, rejected (p.part. of aspernor, despise).
Asperūgo, asper, rough, and -ugo, feminine suffix used in plant-names.
Asperula, diminutive feminine word from asper, rough.

10

Aspidium, ἀσπίδιον, diminutive of ἀσπίς, ίδος, shield, from the form of the indusium.

Asplēnium, ἄσπληνον, name of a plant in Dioscorides, from ἀ, said to be euphonic, and σπλήν, spleen. The *i* in the generally accepted spelling of the generic name is perhaps also euphonic.

Astēr, ἀστήρ, star, as a plant name thought to be *Aster amellus*.

-aster, -astrum, suffix signifying inferior kind, incomplete resemblance, wild, often added to stem of generic name to form designation for a section of the genus. The initial *a* may be changed to *i*, especially when the suffix follows vowel-stems, as in *Sinapistrum* and *Rapistrum*.

Astragalus, ἀστράγαλος, name of a leguminous plant.

Astrantia, name used by l'Ecluse for *Peucedanum Ostruthium*, apparently corrupt form of *Magistrantia* (Germ. *Meisterwürz*, Eng. 'Masterwort'), from *magister*, master.

athamanticus, a, um, of Mount Athamas in Thessaly, or of King Athamas, who first made use of the plant.

Athyrium, said to be derived from ἀθύρω, play, sport, referring to the varying form of the sori.

ātrātus, a, um, clothed in black (*āter*, black).

Atriplex, name of a plant in Pliny; the same word as ἀτράφαξυς.

ātro-, better **ātri-,** black- (*āter, ra, rum*).

ātrofuscus, a, um, from *āter*, black, and *fuscus*, tawny.

Atropa, Ἄτροπος (not to be turned, inflexible), name of one of the three Μοῖραι (*Parcae*) or goddesses of fate.

Atropis, ἀ privative and τρόπις, keel, from the keel-less lemmata.

Aubrieta, called after the artist Claude Aubriet, Tournefort's travelling companion.

aucupārius, a, um, for *aucupātōrius*, used for catching birds (*avis*, bird, and *capio*, catch). Fowlers used the fruit for bait.

aurantiacus, a, um, orange-coloured.

aureus, a, um, golden (*aurum*, gold).

auricomus, a, um, with golden hair, from *aurum*, gold, and *coma* (κόμη), hair of the head, from the numerous yellow flowers, cf. **comōsus.**

Auricula, *auricula ursi* (bear's ear), name, in Clusius, of *Primula Auricula*, so-called from the form of the leaf.

aurigerānus, a, um, of Ariège (*Aurigera*).

aurītus, a, um, with long ears, usually referring to stipules (*auris,* ear).

austrālis, e, southern (*auster*, the south wind, the south).

austriacus, a, um, of Austria.

autumnālis, e, of (flowering in) autumn (*autumnus*, autumn).

avellānus, a, um, of Avella, an Italian town famous for its fruit-trees and nuts. Pliny called the hazel-nut *nux avellana*.

11

Avēna, oats (originally meaning nourishment).

avēnāceus, a, um, oat-like (*avēna*).

aviculāris, e, pertaining to small birds, from *avicula,* diminutive of *avis,* bird. Sparrows and finches feed on the fruits of *Polygonum aviculare*—'quae avide quaeruntur ab aviculis vere, autumno ac hyeme; inde Passerum ager, qui nec serunt, nec in horrea colligunt, quos tamen sic alit eorum Creator' (Linnaeus, *Flora Suecica*).

avium, of birds (gen.pl. of *avis,* bird).

Axyris, said to be derived from ἀξυρής, not cutting, blunt, referring to the bland flavour of the species. Most *Chenopodiaceae,* however, are remarkable for their bland flavour.

Azalea, from ἀζαλέος, dry, of doubtful application.

Azolla, origin obscure.

Babingtonii, of Charles Cardale Babington (1808–95), Professor of Botany at Cambridge, author of *Manual of British Botany, Florà of Cambridgeshire,* and other botanical works.

babylōnicus, a, um, of Babylon. *Salix babylonica* is a native of China.

baccātus, a, um, having a berry-like fruit. (Better spelling, *bacātus,* from *bāca,* berry.)

baccifer, era, erum = bācifer, bearing berries (*bāca,* berry, and *fero,* bear).

Baeothryon, βαιός, small, and θρύον, reed, rush.

baeticus, a, um, of Baetica, ancient province in S. Spain.

Baldellia, called after the Marquis Bartolommeo Bartolini-Baldelli, famous nineteenth-century Italian nobleman.

Ballōta, βαλλωτή, name of *Ballota nigra* in Dioscorides.

balsamifer, a, um, balsam-bearing, *balsamum,* βάλσαμον, balsam, and *fero,* bear.

Balsamināceae, called after *Impatiens Balsamina* (βαλσαμίνη, name in Pseudo-Dioscorides).

Barbarea, dedicated to St Barbara (*Herba Sanctae Barbarae*).

barbarus, a, um, foreign.

barbātus, a, um, bearded (*barba,* beard).

Barkhausenia (Barkhausia), called after Gottlieb Barkhausen, who, in one of his books, gave a list of plants growing in Lippe.

Bartsia, called after Johann Bartsch, Dutch physician, born 1709 in Königsberg, died 1738 in Surinam.

Basitonae, βάσις, base, and τόνος, gerund form of τείνω, stretch, extend. The base of the anther lies against the rostellum.

Batrachia, the species of this section of GERANIUM somewhat resemble in habit *Ranunculus acer* (see *Batrachium*).

12

Batrachium, βατράχιον is the diminutive of βάτραχος, frog, as RANUNCULUS is diminutive of *rāna*. The name is apt, as the species of this section tend to be amphibious. The name, however, was used by Greek authors for several terrestrial species of RANUNCULUS not belonging to our section *Batrachium*.

Batrachoidea, resembling *Batrachia*, as does *Geranium pyrenaicum*.

-batus, suffix used in names of sections of the genus RUBUS (βάτος, name of several plants including RUBUS spp.).

Beccabunga, ultimately from Low German *beckbunge*, from *beck*, stream, and *bunge*, of disputed origin.

Behen, medieval corruption of بهمن, Arabic name of several plants, which was applied to *Silene Cucubalus* and other plants.

Belladonna, originally the Italian name of the plant, meaning beautiful lady, because of a cosmetic prepared from the berries.

Bellis, name of a plant in Pliny.

Berberis, a medieval Latin word of doubtful origin.

Bermudiāna, substantive, used as trivial epithet by Linnaeus. The nomenclature is confused. Our plant is said to be native of Ireland and Pacific North America.

Berteroa, called after Carlo Giuseppe Bertero (1789–1831) of Piedmont, who visited tropical America and died at sea between Tahiti and Chile.

Berula, plant-name in Marcellus Empiricus.

Bēta, the Latin name of the plant.

Betonica, *Vettonica*, the name in Pliny of a medicinal plant growing in the region of Spain called *Vectones* or *Vettones*.

Betula, name of the genus in Pliny.

Betulus, origin obscure. Said to refer to the similarity of the leaves to those of BETULA. *Betulus* (*baetulus*) was the name of a precious stone.

bi-, two-.

bichlorophyllus, a, um, *bi-*, two, χλωρός, green, and φύλλον, leaf, referring to the marked difference in shade between the two leaf surfaces.

Bidens, *bi-*, two, and *dens*, tooth, from the two pappus-bristles. (For the original meanings of *bidens* see Latin dictionaries.)

Biscutella, *bi-*, two, and *scutella*, diminutive of *scutra*, tray (Rich, p. 589), from the form of the fruit.

Bistorta, medieval name, from *bis*, twice, and *tortus*, twisted, referring to the root-stock.

bithỹnicus, a, um, of Bithynia, a province in Asia Minor.

Blackstonia, called after John Blackstone, eighteenth-century English botanical author.

Blattāria, plant-name in Pliny, from *blatta*, cockroach, moth.

Blēchnum, βλῆχνον, name, in Greek authors, of a fern.

Blitum, βλίτον (βλῆτον), name, in Greek and Latin authors, said to be of *Amarantus Blitum.*

Blysmus, origin obscure, thought to be connected with βλύζω, bubble, gush forth.

Bonus-Henrīcus, name perhaps given to distinguish this plant from a poisonous one which was called *Malus-Henricus.* 'Heinrich' of German mythology was troubled with skin disease, and perhaps the plant was used for cutaneous disorders. Other explanations are given (see Hegi, III, p. 219, footnote). See also **Agathophytum.**

Borāgo, medieval Latin *borrāgo*, of doubtful origin; perhaps connected with late Latin *burra,* a shaggy garment, referring to the rough foliage.

Boraphila, badly formed word, from βορέας, the north, and φίλος, friend. Most of the species are arctic, subarctic, or alpine.

boreālis, e, northern.

Boreava, called after Alexander Boreau (1803–75), Professor and Director of the Botanic Garden at Angers; author of *Flore du Centre de la France* and other works.

Botrychium, βοτρύχιον, diminutive of βότρυχος, peduncle of a bunch of grapes.

Botrys, name of a sweet-scented plant in Dioscorides. The word βότρυς ordinarily means a bunch of grapes, which the inflorescence of *Chenopodium botrys* somewhat resembles.

brachy-, short- (βραχύς).

Brachypodium, βραχύς, short, and πόδιον, little foot, from the sub-sessile spikelets.

Brassica, the classical name of several kinds of cabbage.

Brassicella, diminutive of BRASSICA.

brevifimbriātus, a, um, short-fringed (*fimbriātus,* fringed, *fimbriae,* fringe).

brevis, e, short.

Brisēis, Βρισηίς, slave of Achilles.

Britannica (*herba*), name of a plant in Pliny. *Inula Britannica* is not a native of Britain.

Briza, βρίζα, name, in ancient and modern Greek, of rye (*Secale ceriale*).

Bromus, βρόμος, name of kinds of oat in Greek authors.

Brunella, Prunella, word first used by Brunfels, referring to the use of the plant in the inflammation of the throat called by the same name. The spelling with 'B' is the earlier.

brunneus, a, um, brown.

Bryōnia, βρυωνία, name in Dioscorides of a plant called also ἄμπελος μέλαινα (black vine).

Būbōnium, βουβώνιον, name of a plant useful against swellings in the groin (βουβών, ῶνος, groin).

Buda, meaningless word coined by Adanson.

Buffōnia, either called after the celebrated Buffon; or from *būfo*, toad, because the type-species, *B. paniculata,* resembles *Juncus bufonius.*

būfōnius, a, um, having to do with toads (*būfo*, toad).

Bugula, late Latin name of AJUGA spp. Cf. English and French *bugle,* Spanish *bugula,* etc.

bulbifer, a, um, bearing bulbs or, more usually, bulbils (*bulbus,* bulb, and *fero,* bear).

Bulbocastanum, *bulbus,* bulb, and κάστανον, chestnut, referring to the chestnut-like tuber.

Bulliarda, called after P. Bulliard, eighteenth-century French mycologist.

Būnias, βουνιάς, name in Greek authors of a kind of turnip.

Būnium, βούνιον, name in Dioscorides, said to be of *Bunium ferulaceum.* βουνιάς was the name of a kind of turnip.

Būpleurum, βούπλευρος, name of a plant in Nicander. The word is compounded of βοῦς, ox, and πλευρά, rib, side.

Bursa-pastōris, *bursa,* purse, and *pastōris,* gen.sing. of *pastor,* shepherd, from the form of the fruits.

Būtomus, βούτομος, name of a marsh plant in Aristophanes and Theophrastus, probably a monocotyledon with cutting leaves (βοῦς, ox, and τέμνω, cut), thought to be *Carex riparia.*

Buxbaumii, of J. C. Buxbaum (1693–1730) of Saxony, botanical author.

Buxus, πύξος, the name of *B. sempervirens.*

caenōsus, a, um, muddy, dirty, growing in mud (*caenum,* mud, dirt).

caeruleus, a, um, blue.

caesius, a, um, bluish-grey.

caespitōsus, a, um, tufted, forming tussocks (*caespes, itis,* turf, sod).

Cakile, said to be derived from an Arabic word.

Calamagrōstis, name of a plant in Dioscorides, compounded of κάλαμος, reed, and ἄγρωστις (see *Agrostis*).

calamārius, a, um, pertaining to a *calamus.*

calamināris, e, *Thlaspi calaminare* was so called from being found on calamine or zinc spar ($ZnCO_3$).

Calamintha, name of a plant in Nicander and Dioscorides, perhaps from καλός, beautiful, and μίνθη, mint.

Calamus, κάλαμος, reed, cane, in pure Latin *harundo.*

calcarātus, a, um, furnished with a spur (*calcar*).

calcāreus, a, um (pertaining to lime), growing on a soil containing calcium carbonate (*calx, calcis*, limestone, lime).

Calceolus, diminutive of *calceus*, shoe, referring to the form of the lip.

Calcitrapa, word first used by de l'Obel, formed from French *chausse-trape*, meaning caltrop, i.e. a spiked iron ball used for maiming horses. See 'caltrop' in dictionaries.

Calendula, medieval name connected with the calends (*calendae*), the first day of each month in the Roman calendar.

Calepina, origin obscure, perhaps connected with Aleppo, Arabic ﺣﻠﺐ.

Calla, name of a plant in Pliny, ordinarily written *calsa.*

Callitriche, καλλίτριχον, a synonym of ἀδίαντον (see *Adiantum*), from καλός, beautiful, and θρίξ, τριχός, hair.

Callūna, from καλλύνω, to brush. The plant is used for brooms.

Caltha, name in Virgil and Pliny of a plant with yellow flowers, thought to be *Calendula officinalis.*

calvescens, becoming glabrous (pres.part. of *calvesco*, grow bald).

calycīnus, a, um, having a persistent calyx. This is not a proper Greek combination, and is better treated as a Latin word with long *i*.

Calycostegia, from κάλυξ, υκος, calyx, and στέγω, cover. The prophylls enclose the calyx.

Calystegia. See *Calycostegia.*

cambricus, a, um, of Wales (*Cambria*).

Camelīna, a word of obscure origin first used by de l'Obel (see app.).

Campanula, diminutive of *campana*, bell, from the form of the corolla.

campester, tris, tre, pertaining to plains or level country.

canāriensis, e, usually means belonging to the Canary Isles. *Phalaris canariensis* is so called from its use as food for cage-birds, especially canaries. N.B. The Canary Isles were called by the Romans *Insulae Fortunatae*, and one of them was the *Canaria Insula* from its large dogs (*canis*, dog).

candicans, pres.part. of *candico*, be whitish or white.

cānescens, white, lit. becoming white (*cāneo*, be white).

canīnus, a, um, dog-, term applied to wild or useless plants (*canis*, dog).

cannabīnus, a, um, hemp-like. The classical word is *cannabĭnus* (κανναβ-ινος) and means of hemp, hempen.

Cannabis, κάνναβις, *Cannabis sativa*; the word is cognate with 'hemp'.

cantabricus, a, um, belonging to Cantabria, a province in N. Spain.

cantiānus, a, um, of Kent (*Cantium*).

cānus, a, um, white (*cāneo*, be white).

16

capensis, e, modern Latin word meaning native of the Cape of Good Hope. *Impatiens capensis* is a native of N. America.

capillāris, e, hair-like, capillary.

capillātus, a, um, bearing slender hair-like structures (*capillus*, hair of the head).

Capillus-Veneris, hair of Venus, name of a plant, also called *Herba capillaris*, thought to be *Adiantum Capillus-Veneris*.

Capnoïdēs, καπνός (smoke), FUMARIA, and -ιδής, having the form of. The word καπνοειδής meant smoke-coloured.

caprea, wild she-goat.

capreolātus, a, um, having tendrils (*capreolus*, tendril).

Caprifolium, medieval name, said to be translation into Latin of the German word *Geissblatt* (goat-leaf), the name of honeysuckle (*caper*, goat, and *folium*, leaf), but the reverse is more probable.

Capriola, perhaps from *capreolus*, wild goat, which was said to feed on *Cynodon Dactylon*.

Capsella, diminutive of *capsa*, box, from the form of the fruits.

Cardamīne, καρδαμίνη, name of a cress-like herb in Dioscorides, also called καρδαμίς (cf. κάρδαμον, cress).

Cardaminopsis, resembling CARDAMINE (ὄψις, appearance).

Cardamon, κάρδαμον, name, in Greek authors, of *Lepidium sativum*.

Cardāria, word formed from καρδία, heart, referring to the cordate pods.

cardiaca, καρδιακός, to do with the heart (καρδία), because of the medicinal use of the plant.

Carduus, Latin name of various thistle-like plants.

Cārex, name of a plant, probably *Cladium mariscus*.

cāricīnus, a, um, resembling species of the genus CAREX.

Cāricis, gen.sing. of *cārex*.

carīnātus, a, um, keeled (*carīna*, keel).

Carlina, called after Charlemagne, who is said to have used a continental species, *C. acaulis*, to cure a pestilence in his army.

carneus, a, um, flesh-coloured (*caro, carnis*, flesh).

Carōta, καρωτόν, carrot. The word *carota* was used by Apicius for the carrot, which, in Latin, was usually called *pastināca*.

Carpīnus, the Latin name of *C. Betulus*.

Carpobrōtus, καρπός, fruit, and βρωτός, edible.

-carpus, a, um, -fruited (καρπός, fruit).

Carrichtera, called after Bartholomaeus Carrichter, physician to Maximilian II, who published an astrological herbal (*Kreutterbůch*) in 1575.

Carthamus, from the Arabic name of *C. tinctorius*.

carthūsianōrum, of the Carthusian monks. Their first monastery (Grande Chartreuse) still stands at Chartreuse (*Carthusia*) near Grenoble.

Carui (Carvi), a medieval name, probably from an Arabic word for aromatic umbelliferous plants.

Carum, κάρον (καρώ), name of caraway in Dioscorides.

Carvifolia, plant-name (substantive), in old authors, of *Peucedanum Carvifolia*, whose leaves resemble those of *Carum Carui*. The name was unfortunately transferred by Linnaeus to our species of SELINUM, which it does not suit.

Caryolopha, κάρυον, nut, and λόφος, crest, referring to the ring on the nutlets.

caryophyllāceus, a, um, *Orobanche carophyllacea* is so called from its flowers smelling like clove-pinks (*Dianthus Caryophyllus*).

caryophylleus, a, um, resembling a plant belonging to the *Caryophyllaceae*, also clove-coloured, in *Carex caryophyllea*, referring to the bracts.

Caryophyllus, καρυόφυλλον, the name of the clove-tree (*Eugenia caryophyllata*). The meaning, 'nut-leaf', does not apply to this plant, and is an assimilation of the Arabic name. The flowers of *Dianthus Caryophyllus* smell like cloves (see appendix).

Castalia, Κασταλία, the spring of the Muses on Mt. Parnassus.

Castanea, the Latin name of *C. sativa*.

castaneus, a, um, chestnut-coloured.

Catabrōsa, κατάβρωσις, an eating up, devouring, supposed to refer to the lemmata, which are membranous and torn at their tips (see app.).

Catapodium, from κατά, here used in a disparaging sense, and πόδιον, diminutive of πούς, foot, here meaning stalk. The word refers to the sessile or subsessile spikelets (cf. BRACHYPODIUM).

Cataria, old name of a plant (*Herba cataria*) attractive to cats (*catus*, cat).

catenātus, a, um, chained, fettered, p.part. of *catēno*, chain. *Veronica catenata* has long, chain-like inflorescences.

catharticus, a, um, καθαρτικός, purging.

Catōnia, called after M. Porcius Cato, author of *De agri cultura*.

Caucalis, καυκαλίς, name of an umbelliferous plant, perhaps *Tordylium apulum*. See *Anthriscus*.

caulis, stem, used as adjectival termination, -*caulis*, -*e*, -*on*, -stemmed.

cavus, a, um, hollow. The tuber of *Corydalis cava* is usually hollow.

Cedrus, κέδρος, name of several trees with fragrant wood.

Cēlastrāceae, called after the genus of climbing shrubs, CELASTRUS (κήλαστρον in Theophrastus is thought to be ivy).

celerātus, a, um, hastened, quickened (p.part. of *celero*, hasten).

Centaurēa, *centaurēum* and *centauria* in Pliny, κενταύριον, -ειον in Hippocrates, called after Chiron the Centaur, who had wide knowledge of herbs.

Centaurium, same origin as CENTAUREA, q.v.

Centinōde, CENTINŌDIUM, old generic name of *Polygonum aviculare,* from Italian, *centinodia* (*centum,* hundred, and *nōdus,* knot, node).

Centranthus, κέντρον, spur, and ἄνθος, flower, from the spurred corolla.

Centrospermae, fem.pl.adj. (sc. *plantae*), from κέντρον, centre, and σπέρμα, seed, referring to the central placentation.

Centunculus, name of a plant in Pliny. The word also means a small patch, and Dillenius chose the name for this plant because of its insignificance.

Cēpa, *cēpa, caepa, cēpe,* onion.

Cēpaea, called after *Sedum Cepaea* L., a southern species. *Cēpaea* (κηπαία) was the name perhaps of this species (κηπαῖος, cultivated in gardens, κῆπος, garden).

Cephalanthēra, κεφαλή, head, and ἄνθηρα (in botanical usage), anther. The anther is situated on the contracted apex of the column.

Cephalāria, fem.subst. formed from κεφαλή, head, referring to the capitate inflorescence.

-cephalus, a, um, -headed (κεφαλή, head).

Ceramanthē, from κέραμος, jug, and ἄνθη, flowers, referring to the form of the corolla.

cerasifer, a, um, cherry-bearing. The fruits of *Prunus cerasifera* resemble cherries (*cerasus,* cherry, and *fero,* bear).

Cerastium, name first used by Ray, from κεράστης, horned, from the form of the capsule.

Cerasus, κέρασος, the cherry-tree brought from Cerasus, in Pontus, to Italy (see appendix).

cerăto-, horn-, κέρας, horn; the genitive is κέρᾱτος, but the length of the α in compounds is open to doubt.

cerătocarpus, a, um, having a horny fruit (κέρας, κέρᾱτος, horn, and καρπός, fruit).

Cerătochloa, κέρας, ᾱτος, horn, and χλόη, grass, referring to the hornlike lemmata.

Cerătophyllum, κέρας, horn, and φύλλον, leaf, from the texture of the leaves.

cereālis, e, pertaining to Ceres, goddess of agriculture, especially of the cultivation of corn.

Cerefolium. See *Chaerophyllum.*

Cērinthoïdea, called after *Hieracium cerinthoides* (CERINTHE, genus of *Boraginaceae*).

cernuus, a, um, inclined forwards, bowing, nodding.

Cerris, *cerrus,* name in Latin authors of a kind of oak, thought to be *Quercus Cerris.*

Ceterach, said to be derived from a German word meaning itchy, referring to the covering of scales, which resembles a cutaneous eruption.

Chaenomělēs, from χαίνω, gape, and μῆλον, apple or other tree fruit; the fruits were formerly thought to split.

Chaenorrhīnum, word formed on analogy with ANTIRRHINUM, from χαίνω, gape, referring to the open throat of the corolla.

Chaerophyllum, χαιρέφυλλον (apparently from χαίρω, rejoice, and φύλλον, leaf, referring to the ornamental foliage), the name of *Anthriscus Cerefolium*; *caerefolium* in Pliny. Various vernacular derivatives are in use, as English 'chervil'.

Chaixii, *Poa Chaixii,* was called after the Abbé Dominique Chaix (1731–1800), who contributed to Villars' *Histoire des plantes Dauphinoises.*

Chamaecyparis, χαμαί, on the ground, dwarf, and κυπάρισσος, cypress.

Chamaecyparissus, plant-name in Pliny meaning 'ground cypress', from χαμαί, on the ground, and κυπάρισσος, cypress.

Chamaedrys, 'dwarf-oak', χαμαί, on the ground, and δρῦς, oak. χαμαίδρυς was the name in Theophrastus of a plant a span high with oak-like leaves.

Chamaemŏrus, χαμαί, on the ground, and *mŏrus,* mulberry.

Chamaenērion, name used by Gesner, from χαμαί, on the ground, dwarf, and νήριον, oleander.

Chamaepericlymenum, χαμαί, on the ground, dwarf, and *Periclymenum,* q.v. A dwarf plant with honeysuckle-like foliage.

Chamaepitys, χαμαίπιτυς, name of a plant in Theophrastus, etc., from χαμαί, on the ground, and πίτυς, PINUS spp. *Ajuga Chamaepitys* resembles a PINUS seedling and smells resinous.

Chamaetia, χαμαί, on the ground, dwarf, and ἰτέα, willow.

Chamagrōstis, χαμαί, on the ground, dwarf, and AGROSTIS.

Chamomilla, from χαμαίμηλον, name, in Dioscorides, of a plant that smelt of apples, from χαμαί, on the ground, and μῆλον, apple.

Cheiranthus, origin obscure, said to be derived from an Arabic word.

Cheiri, origin obscure.

Chelidonium, χελιδόνιον, name in Theophrastus of *Chelidonium majus,* neuter of χελιδόνιος, of the swallow (χελιδών), perhaps so called from flowering when the swallows arrive and fading away when they leave.

Chelōnēae, called after the N. American genus CHELONE (χελώνη, tortoise, from the form of the upper lip of the corolla).

Chenopodium, name used by l'Ecluse, from χήν, goose, and πόδιον, little foot, referring to the shape of the leaves.

Cherleria, called after J. H. Cherler, son-in-law of G. Bauhin.

chinensis, e, Chinese: *Ch* Sanskritic, not Greek, and pronounced as English *ch*. If treated, according to the International Rules, as a Latin word, the *ch* may be pronounced *c*.

Chionodoxa, χιών, όνος, snow, and δόξα, glory. Boissier discovered *C. Luciliae* flowering among melting snow in western Tmolus.

Chlora, from χλωρός, greenish yellow.

chloranthus, a, um, χλωρός, greenish yellow, and άνθος, flower.

Chlorideae, called after the tropical genus CHLORIS (χλῶρις, greenness; Χλῶρις, goddess of flowers).

chondrospermus, a, um, χονδρός, granular, coarse, and σπέρμα, seed. The seeds of *Montia chondrosperma* are covered with tubercles. χονδρός also means gristle, cartilage.

chordorhizus, a, um, χορδή, string, and ρίζα, root, referring to the long, slender, creeping shoots.

chori-, separate (χωρίς, separately, apart).

Chorispora, χόριον, membrane, and σπόρος, seed; referring to the winged seeds.

Chronosemium, from χρόνος, time, and σημεῖον, flag. In this section the persistent standard enlarges and folds over the fruit.

Chrysanthemum, name in Dioscorides of *C. coronarium*, from χρῡσός, gold, and άνθεμον, flower.

chryso-, gold- (χρῡσός, gold).

Chrysocoma, χρῡσός, gold, and κομή, hair. The stems are terminated by yellow inflorescences (cf. *auricomus*).

Chrysosplenium, from χρῡσός, gold, and σπλήν, spleen, because of its colour, and its use in diseases of the spleen.

Cicendia, name invented by Adanson, probably meaningless.

Cicerbita, *Cicharba*, in Marcellus Empiricus, *De Medicamentis*. *Cicerbita* is modern Italian for *Sonchus oleraceus*.

Cichorium, κιχώριον, plant-name in Theophrastus and Dioscorides. Other forms of the word occur. The word 'succory', used in floras and dictionaries, is a corruption of *cicoree* = chicory.

Cicuta, the Latin name of *Conium maculatum*.

cicutarius, a, um, resembling CICUTA.

Cineraria, name in herbals of *Senecio Cineraria*, so called from its grey leaves (*cinis, eris*, ashes).

cinereus, a, um, ash-coloured (*cinis, eris*, ashes).

Circaea, name in Pliny of a plant used as a charm, and called after Circe. See *lutetianus*.

circinātus, a, um, circular (p.part. of *circino*, to make circular).

Cirsium, κίρσιον, κρίσσιον, name, in Dioscorides, of a plant, thought to be *Carduus pycnocephalus*, used for enlargement of veins (κιρσός).

Cistāceae, called after the Mediterranean genus CISTUS, species of which were called κίσθος by Greek authors.

Cladium, κλάδιον, diminutive of κλάδος, branch. Application to the genus obscure.

clandestīnus, a, um, secret, hidden. See LATHRAEA.

clāvātus, a, um, clavate, club-shaped (*clāva*, club, Rich, p. 172).

clāviculātus, a, um, having tendrils (*clāvicula*, vine-tendril).

Claytonia, called after John Clayton, American physician and botanist.

Clēmatis, κληματίς, name of several climbing plants (diminutive of κλῆμα, twig, vine-twig).

Clēmatītis, κληματῖτις, with long, climbing branches, name of a kind of ἀριστολοχεία in Dioscorides.

Clīnopodium, κλινοπόδιον, name of a plant in Dioscorides, diminutive of κλινόπους, from κλίνη, bed, and πόδιον, little foot. The inflorescences resemble the knobs on the feet of beds. See Rich, p. 178.

Clōstērostȳlēs, κλωστήρ, ῆρος, spindle, and στῦλος, originally meaning pillar, used in botany for style.

Clypeola, better spelt **Clip-,** fem.diminutive of *clipeus*, shield, from the form of the fruit.

Cnīcus, κνῆκος, name of *Carthamus tinctorius* L., a thistle-like plant used for dyeing.

coaetāneus, a, um, coeval (late Latin, *co-*, short form of *com-* (*cum*, with), and *aetāneus*, from *aetas*, age), referring to the simultaneous appearance of leaves and inflorescence.

coccineus, a, um, scarlet, from *coccum* (κόκκος), the name of the gall, formerly thought to be the fruit (κόκκος) of *Quercus coccifera*, used as a scarlet dye.

Cochleāria, from *cochlear* (*coclear*), spoon, referring to the radical leaves of *C. officinalis*.

Cōdōnoprasum, from κώδων, ωνος, mouth of a trumpet, bell, and πράσον, leek.

Coeloglossum, from κοῖλος, hollow, and γλῶσσα, tongue, referring to the chamber at the base of the lip.

-cola (substantive), inhabitant of (*colo*, inhabit, *incola*, inhabitant).

Colchicum, κολχικόν, name of *Colchicum speciosum* in Dioscorides, perhaps after Colchis, to the east of the Black Sea, famous for poisons.

Coleogētōn, name formed on analogy with POTAMOGETON, the first element being κολεός, sheath. The stipules sheathe the axis.

collīnus, a, um, growing on hills (*collis*, hill).

columbārius, a, um, pertaining to doves, dove-coloured (*columba*, dove).

columbīnus, a, um, pertaining to doves, dove-like (*columba*, dove).

Columnae, of Fabio Colonna (1567–1650) of Naples.

Colutea, κολυτέα, name, in Theophrastus and Dioscorides, of a tree, thought to be *C. arborescens*.

Comarum, κόμαρος, name in Theophrastus of *Arbutus unedo*, to the fruit of which the head of achenes of *Potentilla Comarum* bears some resemblance.

commūnis, e, common.

commūtātus, a, um, changed (p.part. of *commuto*, change) (see app.).

comōsus, a, um, furnished with a *coma*, a word applied, in Botany, to tufts of leaves or hairs, and also to terminal inflorescences (*coma*, κόμη, hairs of the head, foliage).

complānātus, a, um, level with the ground (p.part. of *complano*).

concinnus, a, um, neat and beautiful.

condylōdēs, κονδυλώδης, knuckle-like, knobby, from κόνδυλος, knuckle, and -ώδης, having the form of; referring to the perianth tubercle of *Rumex sanguineus* (*R. condylodes* Bieb.).

confūsus, a, m, confused (with other species).

conglomerātus, a, um, crowded together (p.part. of *conglomero*).

Cōniferae, fem.pl. (sc. *plantae*) of *cōnifer, a, um,* cone-bearing (*cōnus*, cone, and *fero*, bear).

Cōnium, κώνειον, the Greek name of *Conium maculatum*, called in Latin *cicūta*, a word now used for another genus of *Umbelliferae*.

connīvens, connivent, i.e. gradually converging (pres.part. of *cōnīveo*, wink).

cōnōpēa (cōnōpsēa), from κώνωψ, ωπος, gnat, mosquito, from the resemblance of the flower to an insect (cf. our species of OPHRYS).

Cōnopodium, from κῶνος, cone, and πόδιον, little foot, foot of a vase, referring to the shape of the stylopodium.

Conringia, called after Hermann Conring, seventeenth-century Professor of Philosophy, Medicine, and Jurisprudence at Helmstedt.

Consolida, name of a plant in the herbal called *Herbarium Apuleii Platonici* and other medieval works, perhaps from *consolido*, make firm, referring to its use in healing wounds. See SOLIDAGO.

contemptus, a, um, despised (p.part. of *contemno*).

contortus, a, um, twisted (p.part. of *contorqueo*, twist).

Convallāria, old name of this plant, ultimately from *lilium convallium* (lily of the valleys), a plant-name in the Vulgate, Cant. ii. 1.

23

Convolvulus, name of a plant in Pliny, from *convolvo*, roll round, interweave.

Conyza, κόνυζα, plant-name in Theophrastus, etc.

corallīnus, a, um, coral-red (κοράλλιον, red coral). The *i* may perhaps be lengthened in the Latin form of the word.

corallioīdēs = *corallinus*.

Corallorhīza, κοράλλιον, coral, and ῥίζα, root, from the appearance of the rhizome. A better spelling would have been CORALLIORRHIZA.

coriāceus, a, um, thick, leathery, coriaceous, opposite of *membrānaceus* (*corium*, leather).

Coriandrum, κορίαννον, κορίανδρον, name in Theophrastus, etc., of *Coriandrum sativum*, from κόρις, bug, referring to the smell of the fruits, and perhaps ἄνισον, aniseed (*Pimpinella Anisum*).

coriifolius, a, um, with coriaceous leaves (*corium*, leather).

coritānus, a, um, of the *Coritani*, a tribe of ancient Britons who inhabited the eastern Midlands, the type locality of *Ulmus coritana*.

corniculātus, a, um, horned, having a long capsule (*corniculum*, diminutive of *cornus*, horn).

cornubiensis, e, of Cornwall (*Cornubia*).

Cornus, the Latin name of *Cornus mas*.

cornūtus, a, um, horned.

Corōnāria, word originally meaning female maker or vendor of crowns or garlands (*corōna*, crown), but employed in herbals as a translation of the feminine of the adjective στεφανωτικός, ή, όν, used for making garlands.

Coronilla, modern diminutive of *corōna*, crown, referring to the inflorescence.

Corōnopūs, κορωνόπους, name of a plant, perhaps *Plantago Coronopus*, in Theophrastus, from κορώνη, crow, and πούς, foot.

Corrigiola, diminutive of *corrigia*, shoe-latchet, rein for a horse.

Corvisartia, called after Jean Nicolas Corvisart Des Marets, physician in ordinary to Napoleon Bonaparte.

Corydal(l)is, first used as a plant-name by Durante. κορυδαλλίς (κορυδός) is the Crested Lark (*Galerida cristata*).

Corylus, the Latin name of the hazel.

Corynēphorus, κορυνηφόρος, club-bearing, from the clubbed awn (κορύνη, club, and -φόρος, bearing, φέρω, bear).

cotinifolius, a, um, with leaves resembling those of COTINUS, a genus of *Anacardiaceae* (κότινος was the Greek name for the wild olive).

Cotoneaster, word invented by Gesner, presumably from *cotonea* for *cydonia*, quince-tree, and *-aster*, suffix meaning wild.

Cotula, medieval name, probably a diminutive of *cota,* Italian name of *Anthemis cota.* The ancient *cotula* (κοτύλη) was a graduated liquid measure used especially for medicines.

Cotylēdon, name of a plant (κοτυληδών), probably *Cotyledon Umbilicus* L. (now called *Umbilicus pendulinus* or *U. rupestris*). The word means any cup-shaped hollow or cavity and refers to the peltate leaves, which are depressed in the centre.

Cracca, name of a leguminous plant in Pliny (see appendix).

Crambē, κράμβη, general name for cabbage-like plants.

Crassula, feminine (sc. *planta*) of the diminutive of *crassus, a, um,* thick, fat 'succulent little plant'.

Crataegus, κραταιγός, name in Theophrastus of *Crataegus heldreichii.*

Crēpis, κρηπίς, name in Theophrastus of *Picris echioides* (perhaps the same word as κρηπίς, shoe).

crēticus, a, um, of the island of Crete.

crīnītus, a, um, covered with hairs (p.part. of *crinio,* cover with hairs, *crīnis,* hair).

crispus, a, um, curly, with wavy margins.

Crista-galli, cock's comb, from the form of the bracts (*crista,* crest, and gen.sing. of *gallus,* cock).

cristātus, a, um, crested, tufted (*crista,* crest).

Crīthmum, κρῆθμον, the name of *Crithmum maritimum,* perhaps from κριθή, barley. The fruits somewhat resemble barley grains.

crocātus, a, um, saffron-yellow. The juice of *Oenanthe crocata* turns yellow on exposure to the air (κρόκος, saffron).

Crocosmia, κρόκος, saffron, and ὀσμή, smell.

Crocus, κρόκος, saffron, i.e. the dried stigmas of *C. sativus.*

Cruciāta, word first used, in the sense of cross-like, by Dodoens, from *crux, crucis,* cross, referring to the arrangement of the leaves.

Cruciferae, fem.pl.adj. (sc. *plantae*) meaning cross-bearing, referring to the 4 petals (*crux, ucis,* cross, and *fero,* bear).

cruentus, a, um, blood-stained (*cruor,* blood).

Crūs-gallī, the leg (*crus*) of a cock (*gallus*).

Cryptogramme, κρυπτός, hidden, γραμμή, line; from the hidden sori.

Cucubalus, name of a plant in Pliny.

Cucurbitāceae, fem.pl.adj. (sc. *plantae*) formed from CUCURBITA, the name of an American genus, whose species are now widely cultivated in numerous varieties. The *cucurbita* of the Romans was *Lagenaria siceria,* which is a native of the Old World. (See Rich, p. 222.)

cuneātus, a, um, wedge-shaped (*cuneus,* wedge).

Cupressus (κυπάρισσος), the name of *Cupressus sempervirens.*

curti-, *curtus,* short, broken short.

25

Cuscuta, name, in Rufinus, of *C. Epilinum* and other species.

cuspidātus, a, um, ending in a sharp, rigid point (*cuspis,* point of javelin).

Cyanus, κύανος, a dark blue substance; name in Meleager of *Centaurea cyanus.*

Cyclamen, κυκλαμίς, κυκλάμῖνος, names of *Cyclamen graecum* in Theophrastus. (Perhaps from κύκλος, circle, from the circular tuber or the spiral fruiting pedicel.)

Cylactis, perhaps mis-spelt from κυλλός, crooked, deformed, and ἀκτίς, ray; of doubtful application.

Cymbalāria, from *cymbalum* (Rich, p. 231), cymbal, from the form of the leaves.

cymbifolius, a, um, boat-shaped, with incurved leaf margins (*cymba,* κύμβη, boat, and *folium,* leaf).

cynanchicus, a, um, used to cure (original meaning, suffering from) a disorder called κυνάγκη (lit. dog-throttling), a kind of sore throat. See 'quinsy' in dictionaries.

Cynapium, 'dog's celery', word used by Tabernaemontanus, from κύων, κυνός, dog, and *apium,* celery.

Cynodon, κύων, κυνός, dog, and ὀδούς, tooth, from the tooth-like buds of the rhizome.

Cynoglōssum, κυνόγλωσσον, name of plant in Dioscorides, from κύων, κυνός, dog, and γλῶσσα, tongue.

Cynosūrus, κύων, κυνός, dog, and οὐρά, tail, from the form of the panicle.

Cyparissias, name in Pliny of a kind of *tithymalus* (EUPHORBIA), from κυπάρισσος, cypress.

Cypērus, κύπειρος, name of *Cyperus rotundus* and *C. longus.* The form κύπερος also occurs.

Cypripedium, Κύπρις, a name of Aphrodite, and πέδιλον, slipper. See *calceolus.* The word should be CYPRIDOPEDILUM.

Cystopteris, κύστις, bladder, and πτερίς, fern, from the bulging indusia.

Cytisus, name of a plant, κύτισος, probably *Medicago arborea.*

Dabeocia, called after St Dabeoc, a Cambro-British saint who settled in Ireland. Wrongly spelt *Daboecia* by Linnaeus.

Dactylis, name of a kind of grape. Perhaps referring, in our genus, to the form of the spikes.

dactyloīdēs, finger-like, from δάκτυλος, finger, and -ειδής, -like, referring to the palmately divided leaves.

Dactylon, δάκτυλος, finger, from the narrow radiating spikes.

damascēnus, a, um, of Damascus, having the colour of the flowers of *Rosa damascena*.

Damasōnium, *damasōnion,* synonym, in Pliny, of *alisma*.

Danaa, called after J. Peter Martin Dana of Turin (eighteenth century).

Danthōnia, called after Étienne Danthoine, who, in the early nineteenth century, did work on the grasses of Provence.

Daphnē, δάφνη, the name of *Laurus nobilis* (see *Laureola*).

daphnoīdēs, δαφνοειδής, laurel-like.

dasy-, thick-, hairy- (δασύς).

dasyclados, δασύς, shaggy, and κλάδος, twig.

dasyphyllus, a, um, δασύς, thick, and φύλλον, leaf.

Datūra, ultimately from the Sanskrit name of *D. Stramonium*, which is a common Indian weed.

Daucus, *daucus, daucum,* in Latin authors said to be the name of the parsnip and carrot; but δαῦκον in Theophrastus was some other plant.

dēbilis, e, weak, feeble.

deca-, dec-, ten- (δέκα).

dēcipiens, deceiving, deceptive.

dēclīnātus, a, um, bent aside, turned down (p.part. of *dēclīno*).

dēcumbens, creeping below, then ascending (pres.part. of *dēcumbo*, lie down).

Delphīnium, δελφίνιον, name of the genus in Dioscorides. The unopened flower somewhat resembles a dolphin (δελφίς, ῖνος).

deltoīdēs, δελτοειδής, triangular, shaped like the capital Greek letter Δ. Usually refers to leaves, but *Dianthus deltoides* is so called from the marks on the petals.

dēmersus, a, um, submerged (p.part. of *dēmergo*).

dēmissus, a, um, low, downcast, dejected.

dens-leōnis, lion's tooth. See LEONTODON.

densus, a, um, dense, condensed, having short internodes.

Dentāria, from *dentārius, a, um,* pertaining to teeth, referring to the tooth-like scales on the root-stock (*dens, dentis,* tooth).

dēnūdātus, a, um, stripped, laid bare (p.part. of *dēnūdo*, from *nūdus*, naked).

dēpauperātus, a, um, impoverished, having a poor appearance (*dē-*, down, and *paupero*, make poor).

dērelictus, a, um, abandoned, neglected (p.part. of *dērelinquo*, forsake).

Deschampsia, named after the French naturalist M. H. Deschamps, who accompanied the expedition sent in search of the ill-fated La Pérouse.

Descurainia, called after François Descourain or Descurain (1658–1740), pharmacist at Etampes in France.

27

Desmazeria, called after Jean Baptiste Henri Desmazières (1796–1862), who wrote on French grasses and other plants.

Deyeuxia, called after N. Deyeux, celebrated early nineteenth-century French chemist.

Dianthus, διόσ-, of Zeus, and ἄνθος, flower (cf. διόσανθος in Theophrastus).

Diapensia, old name of SANICULA, applied by Linnaeus, for no good reason, to our genus.

dictyo-, netted (δίκτυον, net).

didymus, a, um, didymous, i.e. occurring in pairs, or divided into two equal lobes (δίδυμος, double, twin).

diffūsus, a, um, loosely spreading (p.part. of *diffundo,* spread, scatter).

Digitālis, *digitālis, e,* belonging to the finger, name used by Fuchs; translation of German name *Fingerhut,* thimble.

Digitāria, *digitus,* finger, from the radiating spikes.

Digraphis, meaning obscure. Apparently from δι-, double, and γραφίς, a style for writing on waxen tablets.

Dileptium, perhaps from δίς, double, and λεπτός, slender; said, by Wittstein, to refer to the capsule. Perhaps a poor anagram of LEDIPIUM.

dioicus, a, um, dioecious (δίς, twice, double, and οἶκος, house).

Dioscoreāceae, called after the genus DIOSCOREA, whose tubers are known as yams.

Diōtis, δίς, twice, double, and οὖς, ὠτός, ear, from the two spurs of the corolla. See figure of a *diota* in Rich, p. 242. Cf. OTANTHUS.

Diplotaxis, διπλόος, double, and τάξις, disposition, arrangement, from the bi-seriate seeds.

Diplozygieae, from διπλός (διπλόος), double, and ζυγόν, yoke. In this series, though the secondary ridges are the more conspicuous, yet the primary ones also are often in evidence. See *Haplozygieae.*

Dipsacus, δίψακος, name, in Dioscorides, of *Dipsacus fullonum.* The word meant also a kind of dropsy, and, as a plant-name, probably referred to the accumulation of water in the connate leaf bases.

discerptus, a, um, torn in pieces, referring to the deeply cut leaf margins of *Rubus discerptus* (p.part. of *discerpo*).

discolor (of another colour), showing different colours.

dissectus, a, um, with leaves divided into numerous narrow segments (p.part. of *disseco,* cut in pieces).

distachyus, a, um, having two spikes (δι-, twice, and στάχυς). See *stachyon.*

distichus, a, um, arranged in two rows (δι-, twice, and στίχος, row, line).

diurnus, a, um, of (flowering during) the day.

divīsus, a, um, divided (p.part. of *dīvido,* divide).

dīvulsus, a, um, torn apart, separated violently (p.part. of *dīvello*).

28

dolicho-, long- (δολιχός, long).

dolichostachyus, a, um, from δολιχός, long, and στάχυς, spike, referring to the characteristic long, terminal spike of *Salicornia dolichostachya*.

Doria, perhaps connected with French *doré*, gilded, golden.

Doronicum, origin obscure: said to be derived from an Arabic word.

Dortmanna, called after Dortmann, pharmacist in Groningen, who sent the plant to l'Ecluse.

Draba, δράβη, name of a plant in Dioscorides, thought to be *Cardaria Draba*.

Dracunculus, name, in Pliny, of a plant with a twisted, snake-like root. The name was later applied to *Artemisia Dracunculus*.

Drosera, δροσερός, dewy (δρόσος, dew), referring to the clear, shining, dew-like drops of secretion on the leaf-glands. An old name of the genus was *ros solis* (sun-dew).

Drucei, called after George Claridge Druce (1850–1932), famous British botanist, who wrote various works on our flora and bequeathed his herbarium to Oxford University.

Dryas, δρυάς, tree nymph, from δρῦς, oak, from the oak-like leaves.

Dryopteris, δρυοπτερίς, name in Dioscorides of a fern growing on oaks (δρῦς, oak, and πτερίς, fern).

Dulcamāra, 'bitter-sweet', from *dulcis*, sweet, and *amarus*, bitter; the foliage and berries taste first sweet and then bitter.

dūmālis, e, bushy (*dūmus*, thorny bush).

dūmētōrum, of thickets (gen.pl. of *dūmētum*, thicket).

dumnōniensis, e, of Devonshire (*Dumnōnia*).

dunensis, e, growing on sand dunes.

dūriusculus, a, um, somewhat hard (*dūrus*, hard).

dysentericus, a, um, pertaining to (or curing) dysentery.

eborācensis, e, of York (*Eborācum*).

Ebulus, name, in Pliny and Virgil, of *Sambucus Ebulus*.

echīnātus, a, um, prickly (*echīnus*, ἐχῖνος, sea-urchin, hedgehog).

Echīnella, *echīnus*, hedgehog, from the prickly achenes of some species.

Echinochloa, ἐχῖνος, hedgehog, and χλόη, grass, from the conspicuous awns.

Echīnodorus, ἐχῖνος, hedgehog, and δορός, leather bag. Some American species have heads of fruits resembling little hedgehogs, each fruit terminating in a bristle-like beak.

Echīnops, ἐχῖνόπους was the name, in Greek authors, of a prickly plant, from ἐχῖνος, hedgehog, and πούς, foot. Linnaeus altered the second element of this word to ὄψ (for ὄψις), appearance, in order to suit our genus. It is the heads, not the feet, of ECHINOPS that resemble hedgehogs.

29

Echium, ἔχιον, name in Dioscorides of *Echium lycopsis.*

edūlis, edible (*edo,* eat).

effūsus, a, um, spread out, straggling (p.part. of *effundo*).

Elaeagnāceae, called after the genus ELAEAGNUS, ἐλαίαγνος, a shrub-like plant in Theophrastus (ἐλαία, olive, and ἄγνος, *Vitex agnus-castus*).

Elatinē, ἐλατίνη, plant-name in Dioscorides, said to be of *Kickxia spuria.* The adjective, ἐλάτινος, η, ον, means of the fir-tree, ἐλάτη (*Abies cephalonica*). Name given to the modern genus ELATINE because of the resemblance of *E. Alsinastrum* L., a continental species, to a seedling conifer.

ēlātior, ius, comparative of *ēlātus.*

ēlātus, a, um, high, tall (p.part. of *effero*).

Elisma, variation of ALISMA, perhaps suggested by ἐλίσσω, turn round, because of the orientation of the ovules.

elōdēs, mis-spelling of **helōdēs,** ἑλώδης, marshy, fenny, from ἕλος, marsh, and suffix -ώδης, resembling.

Elymus, ἔλυμος, name of a kind of grain in Hippocrates.

Elytrigia, name, said by its author to be connected with ἔλυτρον, husk, glume, lemma.

ēminens, prominent, lofty (pres.part. of *ēmineo,* stand out, project).

Empetrum, name in Dioscorides of a plant found on rocks (ἐν πέτραις). ἔμπετρος was the name of *Frankenia pulverulenta.*

Enarthrocarpus, ἔναρθρος, jointed, and καρπός, fruit.

Endivia, ultimately from ἔντυβον, the same plant as *intybus,* q.v.

Endymiōn, Ἐνδυμίων, name of Diana's famous lover.

ensifolius, a, um, with sword-like leaves (*ensis,* sword, and *folium,* leaf).

Epeteium, Latin neuter, agreeing with SEDUM, of ἐπέτειος, annual.

epigeios, ἐπίγειος, growing on the earth or on dry land. *Calamagrostis epigeios* grows in drier places than its congeners (ἐπί, on, γέα, γῆ, earth).

epihydrus, a, um, resolved form of ἔπυδρος, Ionic for ἔφυδρος, living on the water (ἐπί, upon, and ὕδωρ, water).

Epilinum, word first used by Gerard, from ἐπί, upon, and λίνον (*linum*), flax.

Epilobium, from Gesner's name of the genus ἴον ἐπὶ λοβόν, 'violet (or other flower) on top of the pod', referring to the insertion of the corolla at the end of the long ovary.

Epimēdium, ἐπιμήδιον, name of a plant in Dioscorides.

Epipactis, ἐπιπακτίς, name in Theophrastus of a plant resembling ἐλλέβορος, whence also called ἐλλεβορίνη. The leaves resemble those of VERATRUM, which was called, in herbals, *Helleborus albus* (ἐλλέβορος λευκός).

Epipōgōn, Epipōgium, ἐπί, on, over, and πώγων, beard, from the lip ('beard' of older botanists) being uppermost.

epipsīlus, a, um, from ἐπί (in composition), somewhat, and ψῖλός, bare, stripped, referring to sparse foliage or sparsely bearded petals.

Ēpīrōtēs, ἠπειρώτης, a dweller on the mainland, or on dry land. Name perhaps given in contrast to *Batrachium.*

Epithymum, name in Dioscorides of *Cuscuta Epithymum,* parasitic on thyme (ἐπί, upon, and θύμος, thyme).

Equisētum, *equisaetum,* name of a plant in Pliny, thought to be *Equisetum arvense (equus,* horse, and *saetum,* bristle, hair).

Ēranthis, ἦρ (ἔαρ), spring, and ἄνθος, flower.

Erēmanthē, ἐρῆμος, solitary, and ἄνθη, flower. The flowers of *Hypericum calycinum* are usually solitary.

Erīca, ἐρείκη, probably a species of ERICA.

erīcētōrum, of heaths (gen.pl. of *erīcētum,* community of ERICA, heath).

Ērigerōn, ἠριγέρων, ὁ, name, in Theophrastus, of a plant, probably *Senecio vulgaris,* from ἦρι, early, and γέρων, old man, referring to the early appearance of the white pappus. The word was used by Pliny as a synonym of *senecio,* q.v. Linnaeus makes ERIGERON a neuter word. Cf. SENECIO.

Erinus, ἔρινος, plant-name in Pseudo-Dioscorides, said to be of *Campanula Erinus,* a common Mediterranean plant.

erio-, woolly (ἔριον, wool).

eriocarpus, a, um, woolly-fruited (ἔριον, wool, and καρπός, fruit).

Eriocaulon, ἔριον, wool, and καυλός, stem, from the woolly scapes of some species.

Eriophorum, wool-bearing, ἔριον, wool, and -φόρος, bearing, φέρω, bear, from the conspicuous perianth bristles, which form wool-like tufts.

eriophorus, a, um, wool-bearing, woolly, ἔριον, wool, and -φόρος, bearing, φέρω, bear.

Erōdium, ἐρωδιός, heron, on analogy with GERANIUM, q.v.

Ērophila, (ἦρ) ἔαρ, spring, and φίλος, loving.

errāticus, a, um, wandering, straying, occurring here and there with no fixed habitat (*erro,* wander).

Ērūca, plant-name in Latin authors (see appendix).

ērūcāgo, from ERUCA, and feminine suffix *-āgo.*

Ērūcāria, fem.adj. from ERUCA.

Ērūcastrum, from ERUCA, a genus of *Cruciferae,* and *-astrum,* q.v.

ērūcifolius, a, um, with leaves like those of *Eruca sativa,* a Mediterranean plant sometimes found as an alien in Britain.

Ervum, name in Latin authors of a leguminous plant (see appendix).

31

Ēryngium, ἠρύγγιον, name, in Theophrastus, of a plant with spinous leaves, thought to be *Eryngium campestre*.

Erysimum, ἐρύσιμον, name of a plant in Theophrastus.

Erythraea, ἐρυθρός, reddish, from the colour of the flowers.

Erythrobalanus, ἐρυθρός, reddish, and βάλανος, acorn (oak), referring to *Quercus rubra*.

Escallonia, called after Escallon, Spanish botanist and traveller.

Eschscholtzia, called after Johann Friedrich Eschscholtz (1793–1831), professor at Dorpat.

esculentus, a, um, good to eat.

Esula, name in Rufinus and de L'Obel probably of *Euphorbia* sp.

-ētōrum, gen.pl. of suffix *-ētum*, denoting plant community, as *quercētum*, oak-wood.

eu-, εὖ, well, completely, prefix denoting the type of a genus or species.

Euclidium, εὖ, well, and κλείω, shut, referring to the indehiscent fruit.

Euōnymus, name of a plant in Pliny, εὐώνυμος in Theophrastus. The word means 'of good name or fame'.

Eupatorium, *eupatoria*, name, in Pliny, of *Agrimonia Eupatoria*, called after Eupator, a surname of Mithridates, king of Pontus.

Euphorbia, called by King Juba II of Mauretania after his physician in ordinary, Euphorbus, who first used the latex of *E. resinifera*, and other North African species, medicinally.

Euphrasia, name in herbals, from εὐφρασία, delight, mirth, from εὖ, well, and φρήν, mind.

euprepēs, es, εὐπρεπής, ές, comely, well-looking.

euro-, broad (εὖρος, breadth, width).

eurocarpus, a, um, with broad fruit (εὐρύς, broad, and καρπός, fruit).

Exaculum, diminutive of EXACUM, name of a genus of *Gentianaceae*. *Exacum* was the name of a plant in Pliny.

exiguus, a, um, small, petty.

eximius, a, um, select, choice, uncommon, excellent.

exōtericus, a, um, ἐξωτερικός, external, exoteric, common, prevailing.

expansus, a, um, spread out, expanded (p.part. of *expando*).

Faba, bean.

fabārius, a, um, bean-like (originally, having to do with beans).

Fāgopyrum, *fāgus*, beech, and πῦρός, wheat. The fruits resemble beech-nuts. The word is a translation of the Dutch *boekweit* (English 'buckwheat').

Fāgus, the Latin name for the beech-tree.

Falcāria, from *falx, falcis*, sickle, referring to the form of the leaf segments. The Latin word *falcarius* meant a sickle-maker.

Falcātula, fem. diminutive of *falcātus*, meaning slightly sickle-shaped, referring to the pod.

falcātus, a, um, sickle-shaped (*falx, falcis*, sickle, Rich, p. 273).

fallax, deceitful, deceptive.

Farfara, *farfarus*, name of a plant in Pliny.

farīnōsus, a, um, mealy (*farīna*, meal, *far*, spelt).

fatuus, a, um, insipid, tasteless.

fennicus, a, um, of Finland.

festālis, e, festal, joyous, gay.

Festūca, name of a weed in Pliny. The word also means a straw.

Fīcāria, used as a plant-name by Brunfels. The root-tubers resemble small figs (*fīcus*, fig).

fīcifolius, a, um, with leaves like those of the fig-tree (*fīcus*).

Fīcoidāceae, emended spelling of *Ficoideae*, name, in Lindley and in Bentham and Hooker, of a family which included MESEMBRIANTHEMUM (FICOIDES, old name of MESEMBRIANTHEMUM, from *fīcus*, fig, and *oīdēs*, resembling, referring to the fruit).

Fīlāgo, medieval name from *fīlum*, thread, cobweb, and -*āgo*, feminine termination often used in plant-names. The word refers to the tomentum.

filicaulis, e, having thread-like stems (*fīlum*, thread, and *caulis*, stem).

filiculoīdēs, name in Pliny of a fern, from *filicula*, diminutive of *filix*, fern, and suffix -(*o*)*īdēs*, -(*o*)ειδής, resembling.

filiformis, e, thread-like, from *fīlum*, thread, and *forma*, shape.

Fīlipendula, *filum*, thread, and *pendulus*, hanging, medieval name referring to the interruptedly tuberous root-fibres of *F. vulgaris* Moench. (*Spiraea Filipendula* L.). The tubers seem to hang on threads.

Filix-fēmina, *filix*, fern, and *fēmina*, woman, 'lady fern', medieval name used by Rufinus and Fuchs. See *filix-mās*.

Filix-mās, *filix*, fern, and *mās*, male; combination first used by Fuchs, meaning a fern with bold 'masculine' appearance, 'male fern'.

fistulōsus, a, um, pipe-like, tubular (*fistula*, tube, reed).

flābellātus, a, um, fan-like (*flābello*, fan).

flagellāris, e, stoloniferous (*flagellum*, whip, vine shoot).

Flammula, a little flame, perhaps referring to the burning taste.

flāvescens, turning yellow.

flāvus, a, um, yellow.

flexilis, e, pliant, flexible (*flecto*, bend).

flexuōsus, a, um, tortuous, winding (*flecto*, bend, curve).

flōribundus, a, um, abounding in flowers (*flōs, ōris*, flower, and *abundus*, abundant).

Flōs-cucūlī, *flōs*, flower, and *cucūlī*, gen.sing. of *cucūlus*, cuckoo, probably so called from flowering when the cuckoo calls.

fluitans, floating, when applied to submerged plants, as *Ranunculus fluitans*, means waving about unsteadily in the water.

Foeniculum, *fēniculum*, fennel (*fēnum*, hay).

foenisicii (better **faen-**), wrong gen.sing. of *faenisicia*, mown hay (*faenum*, hay, and *seco*, cut).

foetidus, a, um, stinking (also spelt *fēt-*, *faet-*).

-folius, a, um, -leaved (*folium*, leaf).

fontānus, a, um, growing in springs or fountains (*fons, fontis*, spring, fountain).

formōsus, a, um, finely formed, beautiful.

foulaënsis, e, of the Island of Foula in the Shetlands.

Frāgāria, *frāga*, ? wild strawberries, in Virgil; *frāgum*, strawberry plant in the *Herbarium* (title of a book) of Apuleius Platonicus.

Frāgāriastrum, wild strawberry. See *-aster, -astrum*.

frāgifer, a, um, strawberry-bearing, from the appearance of the head of fruits. See *Frāgāria*.

fragilis, e, brittle. The twigs of *Salix fragilis* separate very readily.

Frangula, medieval name for *Frangula Alnus*. The twigs of this species are very brittle (*frango*, break).

Frankenia, called after J. Franken, a Swedish botanist.

Fraxinus, the name of *F. excelsior* in Virgil, Ovid, etc. (Note short *i*.)

frīgidus, a, um, cold, growing in cold places.

Fritillāria, *fritillus*, a dice-box (Rich, p. 302), from the form of the perianth.

fruticans = **fruticōsus**.

fruticōsus, a, um, shrubby (*frutex, icis*, shrub).

fruticulōsus, a, um, diminutive of *fruticōsus*.

Fuchsia, named after Leonard Fuchs, author of the famous sixteenth-century herbal *De Historia Stirpium*.

fūcifer, a, um, drone-bearing, from *fucus*, drone, and *fero*, bear.

fūliginōsus, a, um, sooty, black (*fūligo, inis*, soot).

fullōnum, gen.pl. of *fullo*, one who fulls cloth. See *fullo* in Rich, p. 303. The heads of *Dipsacus fullōnum* are still thus used.

fulvus, a, um, deep yellow, reddish yellow, tawny.

Fūmāria, ultimately from medieval Latin, *fūmus terrae*, smoke of the earth, i.e. arising from the ground like smoke. Teutonic names, such as the German *Erdrauch*, are translations of the medieval Latin.

fuscus, a, um, dark, swarthy, tawny.

Gagea, called after Sir Thomas Gage (1781–1820), English botanist.

Galanthus, γάλα, milk; ἄνθος, flower, because of the white perianth.

Gāle, Teutonic word (cf. English *gale*, German *Gagel*) used by J. Bauhin and others as a generic name of *Myrica Gale*.

34

Galēga, origin obscure, perhaps, badly formed, from γάλα, milk, and ἀγωγός, drawing forth, eliciting, because *G. officinalis,* when used as fodder, increases the yield of milk in cows.

Galeobdolon, name of a plant in Pliny, from γαλῆ, contraction of γαλέη, weasel, and βδόλος, stench.

Galeopsis, name of a plant in Pliny, from γαλῆ, weasel, and ὄψις, appearance.

galēriculātus, a, um, adjective formed from *galēriculum,* diminutive of *galērum,* helmet-like skull-cap. See *galerus* in Rich, p. 313. The calyx of *Scutellaria galericulata* resembles this head-dress.

Galinsōga, called after Don Mariano Martinez de Galinsoga, Director of the Madrid Botanic Garden at the end of the eighteenth century.

Galium, γάλιον, name, in Dioscorides, of *G. verum,* from γάλα, milk. This plant is widely used to curdle milk for making cheese.

gallicus, a, um (in modern Latin), French.

Gallii, called after Le Gall, author of *Flore de Morbihan* (1852).

gamo-, united (γάμος, marriage).

Gamochaeta, *gamo-,* united, and χαίτη, long hair, mane; in botanical use, bristle. The pappus-hairs are united at the base.

Gastridium, γαστρίδιον, diminutive of γαστήρ, paunch, from the bulging glumes.

Gaultheria, called after Dr Gaulthier of Quebec.

gemellus, a, um, twin, occurring in pairs.

gemmiparus, producing bulbils, from *gemma* (in botanical Latin), bulbil, and *pario,* bring forth, produce. *Spiranthes gemmipara* is not gemmi-parous. The name was given under a misapprehension. See Godfery, *British Orchidaceae,* p. 95.

genāvensis, e, of Geneva.

genēvensis, e. See *genāvensis.*

geniculātus, a, um, bent abruptly at knee-like joints (*geniculum,* little knee, word used by Pliny for a joint or knot on the stem of a plant).

Genista, name, in Virgil, of a broom-like plant.

Gentiāna, name in Pliny of a plant called after Gentius, an Illyrian king.

Gentianella, diminutive of GENTIANA.

gentīlis, e, foreign, belonging to the same race, well-formed, noble.

Geranium, γεράνιον, plant-name in Dioscorides, from γέρανος, crane, from the beak-like fruit.

Gēum, name, in Pliny, of a plant, possibly *Geum urbanum.*

gibberōsus, a, um, badly hunchbacked (*gibbus*).

gibbus, a, um, with a hump-like protuberance.

gigantēus, a, um, γιγάντειος, huge, gigantic, from γίγας, giant.

Gingidium, γιγγίδιον, name in Dioscorides of *Daucus Gingidium.*

Githāgō, probably from *git, gith,* name in Pliny of *Nigella sativa,* and *-āgō,* common suffix in plant-names, from the similarity of the seeds of the two plants. The seeds of *N. sativa* are used as a condiment.

glaber, bra, brum, without hairs, glabrous.

glaciālis, e, icy, frozen, growing in icy situations (*glacies,* ice).

Gladiolus, diminutive of *gladius,* sword, from the form of the leaves.

glandulifer, a, um, glandular, from *glandulae,* tonsils (diminutive of *glans, glandis,* acorn and other nuts), and *fero,* bear.

Glaucium, *glaucion* (note retention of Greek *ον*), plant-name in Pliny; γλαύκιον in Dioscorides was the juice of *Glaucium corniculatum.*

glaucus, a, um, sea-green, covered with a ' bloom' like a plum or a cabbage leaf (γλαυκός, original meaning staring, shining, of the sea).

Glaux, γλαύξ, name in Dioscorides of *Coronopus squamatus.*

Glēchōma, from γλήχων, Ionic for βλήχων, name, in Dioscorides, of *Mentha Pulegium.*

glomerātus, a, um, gathered into a round mass (*glomus,* ball of yarn, etc., Rich, p. 318).

glomerulans, gathering together into little balls (part. formed from *glomerulus,* diminutive of *glomus,* ball or clew of yarn).

glūtinōsus, a, um, viscous, sticky.

Glyceria, γλυκερός, sweet, because of the sweet grain of *G. fluitans.* See A. Arber, *The Gramineae,* p. 1.

glycyphyllus, a, um, γλυκύς, sweet, and φύλλον, leaf, referring to the taste of the foliage.

Gnaphalium, γναφάλιον, name of a plant with felted leaves (γνάφαλλον, felt).

Goldbachia, called after Karl Ludwig Goldbach (1793–1824), Aulic Councillor in Moscow, writer on CROCUS, and on Russian medicinal plants.

gongylōdēs, turnip-like, from γογγύλη, turnip, and *-ώδης,* q.v.

-gōnus, a, um, -angled, usually applied to structures with obtuse angles. See *-quetrus.* (γωνία, corner, angle.)

Goodyera, called after John Goodyer, a seventeenth-century English botanist.

gothicus, a, um (in botanical Latin), native of Gothland, in Sweden.

gracilis, e, slender.

graecus, a, um, Grecian.

Grāmineae, fem.pl. (sc. *plantae*) of *grāmineus.*

grāmineus, a, um, grass-like (originally meaning grassy, covered with grass). *Grāmen* meant grass and other plants.

grandis, e, full-grown, large, lofty, powerful.

grānulātus, a, um, bearing knots or tubercles resembling grains (*grānum,* grain).

grātianopolitānus, a, um, of Grenoble (*Gratianopolis*), ancient capital of Dauphiné.

Gratioleae, called after the genus GRATIOLA. *G. officinalis* is a well-known European medicinal plant formerly called *Gratia Dei* (Grace of God).

grātus, a, um, pleasing, agreeable.

graveolens, strong-smelling (*gravis,* heavy, harsh, and *oleo,* smell).

Griffithii, of J. E. Griffith of Bangor, Wales, who discovered *Potamogeton Griffithii.*

Grindelia, called after David Hieronymus Grindel (1776–1836) of Dorpat and Riga, botanical author.

Grossulāria, from the French *groseille,* gooseberry.

Guizōtia, called after the historian Guizot.

gummifer, a, um, gum-bearing (*gummi,* gum, and *fero,* bear). Application to *Daucus gummifer* obscure.

Gunnera, called after Bishop J. Ernst Gunner (1718–73), author of *Flora Norvegica.*

guttātus, a, um, spotted (*gutta,* drop).

Guttiferae (sc. *plantae*), drop-bearing (plants). Many exotic members of this family exude drops of oleo-resin (*gutta,* drop, and *fero,* bear).

Gymnadenia, γυμνός, naked, and ἀδήν, ἐνος, gland, referring to the uncovered viscidia.

Gymnocarpium, γυμνός, naked, and καρπός, fruit. The indusium is absent.

Gymnogrammē, γυμνός, naked, and γραμμή, line, from the naked sori.

Gymnospermae, fem.pl.adj. (sc. *plantae*), from γυμνός, naked, and σπέρμα, seed.

-gўnus, a, um, -carpelled, -styled, as *trigynus,* having 3 styles (γῠνή, woman, female).

Gypsophila, γύψος, chalk, and φίλος, friend. Some species are calcicolous.

Habēnāria, presumably from *habēna,* strap, rein, of doubtful application.

halimifolius, a, um, with leaves resembling those of *Atriplex Halimus* (ἅλιμον in Greek authors), a Mediterranean species planted in this country.

Halimiōnē, 'daughter of the sea' (ἅλιμος, to do with the sea, and -ώνη, fem. patronymic suffix). The second *i* on analogy with OBIŌNĒ, q.v.

Halorrhāgidāceae, name of family, from the genus HALORRHĀGIS, from ἅλς, salt, and ῥάξ, ῥᾱγός, grape, so called from species growing near the sea and having fruits resembling grape seeds.

Haloscias, ἅλς, salt, and σκιάς, umbel, from the saline habitat.

Hammarbya, called by O. Kuntze after Linnaeus, whose country seat was at Hammarby. Kuntze, considering the name LINNAEA illegitimate, thought that another genus should be called after Linnaeus, so, when altering the name of *Malaxis paludosa*, he called it HAMMARBYA.

hāmulatus, a, um, furnished with a small hook (*hāmulus*, diminutive, *hāmus*, hook).

Haplozygieae, from ἁπλός (ἁπλόος), single, and ζυγόν, yoke, referring to the ridges on the fruits. In this series the secondary ridges are obscure and the primary ones are conspicuous. See *Diplozygieae*.

hebecaulis, e, *hebes, etis,* dull, sluggish, and *caulis,* stem, referring to the prostrate stem of *Rubus hebecaulis.*

Hecatonia, perhaps from ἑκατόν, a hundred, very many. This section of RANUNCULUS includes, in most floras, only the species *R. sceleratus,* which has very numerous carpels.

Hedera, the Latin name of ivy.

Hēdysareae, called after the genus HEDYSARUM (ἡδύσαρον, name of plant in Dioscorides, first element probably ἡδύς, sweet, second element doubtful).

Helenium, ἑλένιον, name of a plant, perhaps *Inula Helenium.* The genus HELENIUM is American.

Heleocharis, ἕλος, marsh, and χάρις, charm, beauty.

Heleogītōn, word formed on analogy with POTAMOGETON, from ἕλος, marsh, and γείτων, neighbour.

Hēlianthemum, from ἥλιος, sun, and ἄνθεμον, flower.

Hēlianthus, ἥλιος, sun, and ἄνθος, flower, referring to the large heads of *H. annuus,* which resemble the sun and turn towards it.

Helictotrichon, ἑλικτός, twisted, and θρίξ, τρῐχός, hair (awn).

Hēlioscopia, ἡλιοσκόπιον, that which looks at the sun, name of a plant, probably *Euphorbia Helioscopia,* in Dioscorides (ἥλιος, sun, and σκοπέω, look at).

Helix, ἕλιξ, anything that assumes a spiral shape, ivy (ἑλίσσω, turn round).

Helleborīnē, synonym of EPIPACTIS, q.v.

Helleborus, ἑλλέβορος, name of *H. orientalis,* a medicinal plant well-known in the ancient world.

Helminthia, from ἕλμινς, ινθος, worm, referring to the slender fruit.

Helobiae, ἕλος, marsh, and βιόω, live.

Helōdēa, from ἑλώδης, marshy, growing in marshy places.

helōdēs. See *Helōdēa.*

Helosciadium, ἕλος, marsh, and σκιάδειον, umbel.

Helxīnē, ἑλξίνη, name, in Dioscorides, of two plants, said to be *Parietaria officinalis* and *Convolvulus arvensis.*

Hēmerocallis, ἡμεροκαλλές, name, in Greek authors, of a plant, thought to be *Lilium Martagon* (ἡμέρα, day, and κάλλος, beauty).

Hēmisphace, ἡμι-, half, and σφάκος, name of SALVIA sp., perhaps meaning that this section does not properly belong to SALVIA.

Hēmizōnia, ἡμι-, half, and ζώνη, girdle. The achenes are half surrounded by their bracts.

Hēpatica, fem. of adj. ἡπατικός, ή, όν, of the liver (ἧπαρ, ἄτος, liver). The plant was used for hepatic disorders on account of the lobed leaves, which were regarded as bearing the 'signature' of the liver.

Hēraclēum, πανακὲς ἡράκλειον ('all-heal of Hercules'), name of a plant in Theophrastus.

herbāceus, a, um, herbaceous, having no woody portion above the ground. *Salix herbacea* is a shrub. The epithet refers to its low stature.

hercynicus, a, um, of the Hercynian Forest (*Hercynia silva*), a region covering a large area of ancient Germany. The name now means of the Hartz Mountains.

Herminium, ἑρμίν or ἑρμίς, prop, support, bed-post, said to refer to the form of the tuber.

Hermodactylus, ἑρμοδάκτυλον, name in Greek authors of two plants thought to be *Colchicum luteum* and *autumnale.* The word means fingers of Hermes ('Ερμῆς, and δάκτυλος) and is applied to our genus because of its palmate tubers.

Herniāria, word first used by Dodoens, from *hernia*, rupture, for which it was formerly employed.

Hērorchis, ἥρως, hero, and ORCHIS, perhaps referring to *Orchis militaris.*

Hesperis, ἑσπερίς, name in Theophrastus of a plant that flowered in the evening (ἕσπερος, evening).

hetero-, different (ἕτερος).

Heterochlamydeae, fem.pl.adj. (sc. *plantae*), from ἕτερος, different, and χλαμύς, cloak (perianth), i.e. with perianth consisting of calyx and corolla.

heterophyllus, a, um, having leaves of different forms.

Heterosciadeae, from ἕτερος, different, and σκιάς, άδος, umbel, from the irregular umbels.

hex-, hexa-, six- (ἕξ).

hībernus, hībernicus, a, um, Irish.

hībernus, a, um, belonging to winter.

hiemālis, e, of (flowering in) winter (*hiems*, winter). *Eranthis hiemalis* flowers during winter and *Equisetum hiemale* has overwintering stems.

39

Hieracium, name of two plants in Dioscorides, from ἱέραξ, hawk, because this bird was said to use them for diseases of the eye.

Hierochloë, ἱερός, holy, and χλόη, grass. Our species is, in some countries, held sacred to Our Lady.

Himantoglōssum, ἱμάς, ἱμάντος, leather strap, thong, and γλῶσσα, tongue, from the long, narrow lip.

Hippocastanum, word used by Tournefort, from ἵππος, horse, and κάστανον, chestnut, probably meaning chestnut unfit for food (cf. canīnus).

Hippochaete, ἵππος, horse, and χαίτη, mane (bristle), on analogy with EQUISETUM.

Hippocrēpis, ἵππος, horse, and κρηπίς, shoe, from the shape of the joints of the pod. The old botanists called the genus *ferrum equīnum*.

Hippophaë, ἱπποφαές, name of a plant in Theophrastus, apparently from ἵππος, horse, and φάος (φῶς), light.

Hippūris, ἵππουρις, 'horse-tail', both the real thing and EQUISETUM (ἵππος, horse, and οὐρά, tail).

hircīnus, a, um, goat-like, smelling of goats (*hircus*, he-goat).

Hirculus, name of a plant in Pliny, diminutive of *hircus*, he-goat.

Hirschfeldia, called after C. C. L. Hirschfeld, a botanist of Holstein.

hirtus, a, um, hairy.

hispānicus, a, um, Spanish.

Holcus, name, in Pliny, of a kind of grain.

Holoschoenus, ὁλόσχοινος, name of *Holoschoenus vulgaris* in Theophrastus.

holosēricus, a, um, ὁλοσηρικός, all of silk, referring to the covering of long hairs (ὅλος, whole, entire, and σηρικός, silken).

Holosteum, ὁλόστεον, name of a plant in Dioscorides, from ὅλος, whole, entire, and ὀστέον, bone.

homo-, ὁμός, one and the same, agreeing in every detail.

Homogyne, see *homo-* and *-gÿnus*. The styles of the carpellary and hermaphrodite flowers are alike.

homoio-, ὅμοιος, similar, resembling, but not identical (see appendix).

Honckenya, called after G. H. Honckeny, author of *Synopsis plantarum Germaniae*.

Hordeum, the Latin name of barley.

hormīnoīdēs, resembling *Salvia Horminum*, ὁρμῖνον of Greek authors.

Hornungia, called after Ernst Gottfried Hornung (1795–1862), who made a catalogue of scientific papers.

horridus, a, um, bristly, prickly.

hortensis, e, cultivated in gardens (*hortus*, garden).

Hottōnia, called after Peter Hotton (1648–1709), Professor of Botany at Leyden.

humifūsus, a, um, spread out on the ground (*humus*, ground, *fundo, fūsum,* pour, spread).

humilis, e, on the ground, low (*humus,* ground, cf. χαμαί, on the ground).

Humulus, latinized form of a Teutonic word for hops, perhaps of Middle Dutch *Hommel* (cf. modern Danish *humle*).

Hutchinsia, called after Miss Hutchins of Bantry, who died about 1816. She contributed largely to Smith's *English Botany.*

Hyacinthus, ὑάκινθος, name of a youth who was accidentally killed. From his blood sprang the flower of the same name marked with the exclamation '*AI*'. As a plant-name in Greek authors said to be *Scilla bifolia* and *Delphinium Ajacis.* See *Ājācis.*

hybridus, a, um, hybrid. The name has been applied to plants which are certainly not of hybrid origin.

Hydastȳlus, name suggested by Dryander; origin obscure.

Hydatica, ὑδατικός, for ὑδάτινος, wet, watery.

Hydrangea, ὕδωρ, water, and ἄγγος, vessel, from the form of the capsule. The Greek word for water pail was ὑδρία (Rich, p. 342).

Hydrilla, probably an ill-formed diminutive of ὕδρα, water serpent.

hydro-, ὑδρο-, combined form of ὕδωρ, ατος, water.

Hydrocharis, ὕδωρ, water, and χάρις, charm, beauty.

Hydrochloa, ὕδωρ, water, and χλόη, grass.

Hydrocotyle, ὕδωρ, water, and κοτύλη, cup, from the habitat and the form of the leaves.

Hydrolapathum, name of a plant in Pliny, from ὕδωρ, water, and λάπαθον, RUMEX spp.

hydropiper, ὕδωρ, water, and *piper,* pepper, referring to the acrid taste of *Polygonum Hydropiper,* and possibly to the fruits of *Elatine Hydropiper,* which somewhat resemble peppercorns.

hyemalis. See *hiemalis.*

Hymenophyllum, ὑμήν, ένος, thin skin, membrane, and φύλλον, leaf, from the membranous fronds.

Hyoscyamus, ὑοσκύαμος, name in Dioscorides, of species of HYOSCYAMUS, from ὗς, pig, and κύαμος, bean.

Hypericum, ὑπέρεικον, in Dioscorides, name of a plant like ἐρείκη. Different spellings occur. In this country the *i* is wrongly treated as short.

Hypochoeris, ὑποχοιρίς, name of a plant in Theophrastus. The origin of the word is doubtful (see appendix).

Hypoglōttis, pre-Linnaean name applied to several species of ASTRAGALUS, from ὑπό, beneath, and γλωττίς, mouth of the wind-pipe (*rima glottidis*), referring to the groove on the pod.

hypophēgea, ὑπό, beneath, and φηγός, *Quercus Vallonia,* wrongly supposed to mean *fāgus,* with which the word is cognate. *Monotropa hypophēgea* often grows in beech-woods.

hypopitys, ὑπό, under, and πίτυς, PINUS spp., from growing in coniferous woods.

hȳssōpifolius, a, um, with leaves resembling those of HYSSŌPUS.

Hȳssōpus, ὕσσωπος, plant-name, of Hebrew origin, in Latin and Greek authors and in the Septuagint and Vulgate.

Hystrix, ὕστριξ, ιχος, a porcupine. The corm of *Isoetes Hystrix* is clothed with spinescent, dark, horny leaf bases.

Ibēris, ἰβηρίς, name of a plant in Dioscorides.

īdaeus, a, um, belonging to Mount Ida. See *Vītis-Idaea.*

Īdēobatus, from *Rubus idaeus* and *-batus* (q.v.).

-īdēs or **-oīdēs,** -like (-(o)ειδής, from εἶδος, shape).

Īlex, the Latin name of *Quercus Ilex.*

Illecebrum, from *illecebra,* name of a plant in Pliny. The word means enticement (*illicio,* allure, entice).

ilvensis, e, of the Island of Elba (*Ilva*), where, however, *Woodsia i.* does not grow. The name was given in error.

imberbis, e, beardless (*in-* negative prefix, and *barba,* beard).

imbricātus, a, um (of leaves, scales, etc.), overlapping (*imbrex, icis,* tile, *imber,* shower of rain).

Impatiens, adjective meaning impatient, referring, in the genus IM-PATIENS, to the sudden dehiscence of the capsule when touched. Used, for the same reason, as a trivial name for a species of CARDAMINE.

Imperāti, of Ferrante Imperate, sixteenth-century Italian naturalist.

Imperātōria, from *Imperātor, ōris,* commander, master, referring to the much-vaunted medicinal properties of *Peucedanum Ostruthium* ('master wort').

incānus, a, um, white (*cāneo,* be white).

incarnātus, a, um, made of flesh, referring to the colour (*incarno,* make flesh).

incomparābilis, e, that cannot be equalled, incomparable.

incubāceus, a, um, lying close on the ground (*incubo,* sit upon eggs).

incurvus, a, um, bent, curved.

inermis, e, unarmed, without thorns, without awns.

infestus, a, um, hostile, troublesome, hence invading cultivated ground and becoming a weed.

inflexus, a, um, p.part. of *inflecto,* bend, bow, curve. The inflorescence branches of *Juncus inflexus* tend to curve inwards, those of *J. effusus* to spread.

42

infundibulum, a funnel, a hopper for grain. See Rich, p. 352 (*infundo*, pour in).

innātus, a, um, inborn, inherent, natural, applied to *Corallorhiza innata* for *adnatus*, the spur being adnate to the ovary.

inodōrus, a, um, without smell.

insitīcius, a, um, grafted.

insubricus, a, um, for *insuber, bris,* belonging to *Insubria*, a district in the neighbourhood of Milan.

intactus, a, um, untouched, uninjured (see appendix).

integer, gra, grum, entire.

Intybus, name, in Virgil, etc., of a plant, probably a cultivated CICH-ORIUM. Various forms of the word occur.

Inula, name, in Pliny and Columella, of a plant synonymous with ἐλένιόν (see HELENIUM).

inundātus, a, um, flooded, growing in wet places (*inundo*, flow over).

involūtus, a, um, obscure, intricate.

īricus, a, um, Irish.

īrio, name, in Pliny, of a plant called σισύμβριον by the Greeks. (See *Sisymbrium*.)

Iris, Ἶρις, rainbow, name of the plant.

irrigātus, a, um, flooded, growing in wet situations (p.part. of *irrigo*, to water, irrigate).

Īsatis, ἰσάτις, name of a dye-plant, probably woad, in Hippocrates and Dioscorides. *Glastum* is the Latin name of woad.

Ischaemum, name, in Pliny, of a styptic plant, from ἴσχω, hold, stop, and αἷμα, blood.

Isnardia, called after A. Danty d'Isnard, eighteenth-century Professor of Botany in Paris.

īso-, equal (ἴσος).

Isoëtes, name, in Pliny, of a plant synonymous with *aïzoon* and *sempervivum* (q.v.). The Greek word ἰσοετής (ἰσοέτηρος) meant equal in years (ἴσος, equal, and ἔτος, year), but Pliny evidently meant evergreen. Pliny's word is neuter, hence the termination *ēs* (ες). His plant was probably a member of the *Crassulaceae*.

Isolepis, ἴσος, equal, and λεπίς, scale. The lower glumes are not strikingly different from the upper ones.

-ītēs, Greek suffix -ίτης, connected with or belonging to.

Iva, probably Romance name, current in Switzerland, of *Achillea tomentosa*, applied to our genus by Linnaeus, perhaps because of its aroma; but the word was used, as a plant name, by Rufinus.

jacea, medieval name of CENTAUREA spp.

43

Jacōbaea, a former name of *Senecio Jacobaea.* This and allied species were called, in herbals, *Herba Sancti Jacōbi* and similar names, from being supposed to begin to flower on St James's Day (25 July).

japonicus, a, um, modern Latin adjective meaning Japanese.

Jasiōnē, ἰασιώνη, name of *Calycostegia sepium* (? ἴασις, healing).

jubātus, a, um, having a mane or crest (*juba*), referring to the numerous long awns of *Hordeum jubatum.*

Juglans (*Jovis glans*, Jupiter's nut), walnut.

Juncāgināceae, from JUNCĀGO, Tournefort's name for TRIGLOCHIN (JUNCUS, and fem. suffix *-āgo*).

junceus, a, um, rush-like (JUNCUS).

Juncus, the Latin name for rushes, i.e. JUNCUS spp. and similar plants.

Jūniperus, the Latin name of JUNIPERUS spp.

Kalī, ultimately from an Arabic word first meaning the calcined ashes of SALSOLA and SALICORNIA spp., afterwards transferred to the plants themselves. Our word alkali is the same word with the Arabic article prefixed.

Kalmia, called after Pehr Kalm (1715–79), Finnish author.

Kentranthus, see CENTRANTHUS.

Kērosphaereae, κηρός, bees-wax, and σφαῖρα, ball, sphere. The pollen coheres in masses of a waxy texture.

Kickxia, called after the Belgian botanist J. Kickx, who died in 1831.

Knautia, called after Christian Knaut, doctor at Halle (1654–1716), author of *Methodus plantarum genuina.*

Kobresia (Cobresia), called after Paul von Cobres (1747–1823), whose collections were bought by the Bavarian Academy of Science.

Koeleria, called after G. L. Koeler, a German writer on grasses.

Koenigia, called after Johann Gerhard Koenig, pupil of Linnaeus, traveller, medical missionary and plant collector.

Kohlrauschia, called after H. Kohlrausch, sedulous lady botanist, who contributed to the flora of Berlin during the early nineteenth century.

Labiātae, fem.pl.adj. (sc. *plantae*), meaning lipped, referring to the 2-lipped corolla (*labium*, lip).

Laburnum, name of the plant in Pliny.

laciniātus, a, um, fringed, cut into deep irregular segments (*lacinia*, lappet, edge of a garment).

lacistophyllus, a, um, having torn leaves (λακιστός, torn, φύλλον, leaf).

lacteus, a, um, milk-coloured (*lac, lactis*, milk).

Lactūca, the Latin name for lettuce (*lac*, milk, from the white juice).

Lactūcella, diminutive of LACTUCA.

lacuster, tris, tre, associated with lakes or ponds (*lacus*, lake). A masculine in *-tris* is in use.

Lādanum, the Latin name of the exudate of *Cistus creticus* (*lāda*).

laetivirens, with bright green foliage (*laetus*, joyful, *vireo*, be green).

laevigātus, laevis. See *lēvigatus, lēvis*.

Lagarosīphōn, λαγαρός, thin, narrow, and σίφων, tube, pipe, referring to the long, slender tube of the carpellary flowers.

lagōpīnus, a, um, like a hare's paw (λαγώπους).

lagōpūs, a, um, λαγώπους, rough-footed like a hare; as a substantive name, in Dioscorides, of *Trifolium arvense* (λαγώς, hare, and πούς, foot).

Lagūrus, λαγώς, hare, and οὐρά, tail, from the soft, hairy inflorescence.

Lāmiopsis, from LAMIUM and ὄψις, appearance.

Lāmiotypus, from LAMIUM, and τύπος, original pattern, model, type, here referring to the genus LAMIUM.

Lāmium, name of *L. maculatum* in Pliny. The word is thought, by some, to be connected with λαιμός, throat, gullet.

lamprocarpus, a, um, λαμπρός, shining, and καρπός, fruit.

lamprospermus, a, um, λαμπρός, shining, and σπέρμα, seed.

lānātus, a, um, woolly (*lāna*, wool).

Landra, name of *Raphanus Landra* in north Italy, where it is eaten as a salad.

Lantāna, a late Latin word for *viburnum*, cf. Italian *lantanna, lantana*.

lānūginōsus, a, um, woolly (*lānūgo*, woolly substance, *lāna*, wool).

lapathifolius, a, um, sorrel-leaved, from *lapathum* (λάπαθον) RUMEX spp., and *folium*, leaf.

Lapathum. See *lapathifolius*.

Lappa, Latin name for a bur.

lappāceus, a, um, resembling a bur (*lappa*).

lappōnum, of the Lapps (*Lappones*), who use *Salix lapponum* with *Betula nana* as fuel for their never extinguished fires. 'Cum Betula nana in Alpibus alit focum perennem Lapponum' (Linnaeus, *Flora Suecica*).

Lappula, diminutive of *lappa*, bur, referring to the nutlets.

Lapsana, λαμψάνη, or λαψάνη, name in Dioscorides of a potherb, perhaps *Raphanus Raphanistrum*, which is still called *lampsana* in Apulia.

Larbrea, called after Antoine l'Arbre, author of a Flora of Auvergne.

Laricio, Italian name of *Pinus nigra* and *P. Pinaster*.

Larix, the name, in Latin authors, of *Larix decidua*.

lasio-, shaggy- (λάσιος, shaggy).

lasiolaenus, a, um, λάσιος, shaggy, and λαῖνα, for χλαῖνα, cloak, wrapper, from the hairy involucre.

45

Lastrea, called after Charles Jean Louis Delastre, nineteenth-century botanical author, who wrote on the flora of the Department of Vienna.

Lathraea, λαθραῖος, hidden, a translation of Tournefort's name, *Clandestīna,* of *L. Clandestina,* a purple-flowered European species, naturalized in botanic gardens and other localities. The name refers to the lowly habit of the plant.

Lathyris, name of *Euphorbia lathyris*

Lathyrus, λαθύρος, name of *Lathyrus sativus.*

lāti-, broad- (*lātus, a, um*).

latobrigōrum, gen.pl. of *Latobrigī,* name, in Caesar, of a Gallic people, who lived probably in the Rhineland.

Laureola, diminutive of *laurea* (sc. *corona*), a garland of laurel. The evergreen leaves of *Daphne Laureola,* aggregated towards the end of the stem, resemble such a garland.

Laurocerasus, combination of *laurus,* laurel, and *cerasus,* cherry.

Lavatēra, called after the Swiss physician Lavater, friend of Tournefort.

laxi-, laxus, a, um, loose, open.

Lēdum, λῆδον, the name of species of CISTUS yielding lādanum.

Leersia, called after J. D. Leers, a German botanist.

Legousia, derivation unknown.

Legūminōsae, fem.pl.adj. (sc. *plantae*) from *legūmen, inis,* general name of leguminous plants, especially the bean, from *lego,* pluck, gather.

Lemna, λέμνα, name in Theophrastus of a water plant, probably a species of CALLITRICHE.

lendiger, a, um, *lens, tis,* nit, and *gero,* bear, from the appearance of the spikelets.

Lentibulāriāceae, fem.pl.adj. (sc. *plantae*), formed from LENTIBULARIA, Gesner's name of UTRICULARIA, probably from mis-spelling of *lenticula,* diminutive of *lens, tis,* lentil, referring to the bladders. Hegi suggests *lens,* and *tubula,* little tube.

lentīginōsus, a, um, freckled (*lentīgo, inis,* freckle).

Leontodon, λέων, οντος, lion, and ὀδούς, όντος, tooth. The word, made by Linnaeus, is a translation of *Dent de lion* (Dandelion), the French name of *Taraxacum officinale,* included by Linnaeus in LEONTODON, and refers to the form of the leaf of that plant.

Leōnūrus (properly **Leontūrus**), λέων, οντος, lion, and οὐρά, tail, referring to the long inflorescence.

Lepia, probably a shortened form of the word *lepidium.*

Lepidium, λεπίδιον, name of a plant in Dioscorides (diminutive of λεπίς, scale). The fruits of the genus LEPIDIUM are scale-like.

lepido-, scaly- (λεπίς, ίδος, scale).

Lepidobalanus, λεπίς, ίδος, scale, and βάλανος, acorn, referring to the numerous appressed scales on the cupule.

lepidocarpus, a, um, λεπίς, ίδος, scale, and καρπός, fruit. The perigynia of *Carex lepidocarpa* are flatter and more scale-like than those of *C. tumidicarpa*.

lepidus, a, um, pleasant, charming, neat.

Lepigonum, probably badly formed word from λεπίς, ίδος, scale, and γόνυ, ατος, knee, referring to the stipule-clad nodes. Wittstein suggests γόνος, seed, as the second element. The seeds of *Spergularia marginata* are winged and scale-like.

leporīnus, a, um, pertaining to a hare. The spikes of *Carex leporina* L. somewhat resemble hares' paws (*lepus, oris*, hare).

leptochīlus, a, um, having a slender lip (λεπτός, slender, and χεῖλος, lip).

leptoclados, with slender shoots (λεπτός, slender, and κλάδος, branch).

leptolepis, λεπτός, small, fine, thin, and λεπίς, scale.

leptophyllus, a, um, with slender leaves (λεπτός, slender, and φύλλον, leaf).

Leptūrus, λεπτός, slender, and οὐρά, tail, from the slender, tail-like spikes.

Leucanthemum, name, cited in Dioscorides as synonymous with ἀνθεμίς and παρθένιον, from λευκός, white, and ἄνθεμον, flower.

Leucē, a name of *Populus alba* (λεύκη).

leuco-, white- (λευκός).

Leucojum (Leucoium), λευκόϊον (λευκὸν ἴον, white-violet), name applied by the Greeks to several plants. The word ἴον was used for various plants, especially ones with sweet-scented flowers.

Leucorchis, λευκός, white, and ORCHIS.

lēvigātus, a, um (often mis-spelt *laevigātus*), smooth (p.part. of *lēvigo*, make smooth).

lēvis, e (often mis-spelt *laevis*), smooth.

Leycesteria, called after William Leycester, a judge in Bengal.

Libanōtis, λιβανωτίς, name of strongly scented plants (λιβανωτός, incense).

Ligusticum, name of a plant growing in Liguria (λιγυστικός, Ligurian).

Ligustrum, name, in Virgil, of a plant with white flowers.

Lilium, the name, in Virgil, etc., probably for species of LILIUM and other genera.

Limnanthemum, λίμνη, pond, and ἄνθεμον, flower.

Limōnium, λειμώνιον, name of a plant in Dioscorides, from λειμών, meadow.

Limōsella, fem.dim. (sc. *planta*) of *limōsus*, muddy, from the habitat.

limōsus, a, um, growing in muddy places (*limus*, mud).

Lināria, name, in Mattheus Sylvaticus, of *Linaria vulgaris*, referring to the resemblance of the leaves to those of flax (*linum*).

47

Lingua, name of a plant in Pliny. The word means tongue, and was applied to several plants with tongue-like leaves, e.g. *Lingua canis*, *L. būbula*.

lingulātus, a, um, tongue-shaped (*lingua*, tongue).

līnicola, substantive, living in fields of flax (*līnum*, flax, and *colo*, inhabit).

Linnaea, called after Linnaeus at his own request. See HAMMARBYA.

Linosyris, name first used by de l'Obel, from λίνον, flax, and ὄσυρις, plant-name in Dioscorides. The υ is short, and the word is wrongly accented in Hooker's Flora.

Līnum, the Latin word for flax; the Greek is λίνον.

līolaenus, a, um, λεῖος, smooth, and λαῖνα, for χλαῖνα, cloak, wrapper, from the glabrous involucre.

Liparis, λιπαρός, greasy, from the surface of the leaves.

Listĕra, called after Dr Martin Lister, physician in ordinary to Queen Anne, and one of the first investigators of fossils.

Lithospermum, λιθόσπερμον, name of *L. officinale* in Dioscorides, from λίθος, stone, and σπέρμα, seed. The nutlets in *Boraginaceae* and *Labiatae* were regarded as naked seeds.

litorālis, e, belonging to the shore (*lītus*, shore).

Lītorella, *lītus*, shore, from the habitat.

Lloydia, called after Edward Lloyd, sometime Keeper of the Ashmolean Museum, Oxford. He discovered *L. serotina* in Wales.

Lobelia, called after Matthias de l'Obel, Flemish botanist (1538–1616), author of *Plantarum seu Stirpium Historia*.

-lobium, -lobion, with pod-like fruit of the form designated by the first part of the word (see below).

Lobulāria, fem.dim. word formed from λοβός, pod, referring to the small fruits.

-lobus, a, um, -lobed (λοβός, lobe, pod-like fruit). This suffix often refers to pod-like fruits or to cotyledons.

lochabrensis, e, of Lochaber, district of Inverness-shire.

Lōcusta, locust, old name of *Valerianella locusta* (see appendix).

Loeselii, of Johann Loesel, author of *Flora Prussica* (1703).

Lōganiāceae, called after the genus LOGANIA, which was named in honour of J. Logan, author of *Experimenta de plantarum generatione*, born in Ireland, 1674, went to America with Penn, became Governor of Pennsylvania, died 1751.

loganobaccus, the loganberry was produced by Judge J. H. Logan, of Santa Cruz, California (*baccus*, for *bāca*, berry).

Loiseleuria, called after Loiseleur-Deslongchamps, a French botanist.

loliāceus, a, um, resembling LOLIUM.

Lolium, name, in Virgil, of a troublesome weed (see appendix).

Lōmāria, from λῶμα, ατος, hem, border of a garment, referring to the marginal sori.

Lonchītis, λογχῖτις, the name of a plant in Dioscorides (λόγχη, spear-head, spear).

lonchophyllus, a, um, with spear-like leaves (λόγχη, spear, and φύλλον, leaf).

longus, a, um, tall, long.

Lōnicēra, called after Adam Lonitzer (1528–86), German physician and botanical author.

Lōranthāceae, called after the genus LORANTHUS, which has strap-like tepals (λῶρον, strap, thong, and ἄνθος, flower).

Lōroglōssum, λῶρον, thong, and γλῶσσα, tongue, from the long, narrow lip.

Lōtus, *lōtos* (λωτός), the name used for several plants.

lūcens, shining (*lūceo,* shine).

lūcidus, a, um, shining (*lūceo,* shine).

Ludwigia, called after C. G. Ludwig, eighteenth-century professor at Leipzig.

Lūnāria, *lūna,* moon, to which the persistent, glistening, silver repla of LUNARIA are likened. *Botrychium Lunaria* was called *Lunaria minor* in Fuchs and Mattioli. For a figure of this plant as a 'lunar herb', see Arber, p. 257.

Lupīnus, the Latin name of *L. albus,* said to be connected with *lupus,* wolf.

lupulīnus, a, um, hop-like. See LUPULUS. The name was first applied to species of TRIFOLIUM with hop-like inflorescences (e.g. *T. campestre*), and afterwards transferred to the plant now called *Medicago lupulina,* which it does not suit.

Lupulus, medieval name used by Brunfels. Possibly connected with *lupus,* wolf. *Lupus salictārius* is the name in Pliny of a twining plant, probably *Humulus Lupulus,* injurious to willows. The hop is commonly seen climbing on the willows in the continental *Auenwälder,* like a wolf overwhelming its prey.

Luronium, said, by Rafinesque, to be an ancient name for ALISMA.

lūsitānicus, a, um, for *lūsitānus,* Portuguese.

lutārius, a, um, living on mud (*lutum,* mud).

lūteolus, a, um, diminutive of *luteus.*

lutetiānus, a, um, of Paris (*Lutetia*). *Circaea lutetiana* was so called because the *Parisian* botanists held it to be the *Circaea* of the ancients.

lūteus, a, um, yellow (from *lūtum,* name of *Reseda luteola,* which yields a yellow dye).

Luzula, probably from the Italian name *Lucciola,* perhaps from *luceo,* shine.

Lychnis, λυχνίς, name of a plant in Theophrastus, from λύχνος, lamp. The leaves of λυχνίς στεφανωτική (probably *Coronaria tomentosa*) were used as wicks.

Lychnitis, name, in Pliny, of a plant used for lamp-wicks (λύχνος, lamp).

Lycium, λύκιον, name of a thorny tree found in Lycia, perhaps a species of RHAMNUS, probably *R. infectoria.*

Lycopersicum, λύκος, wolf, and περσικόν, peach. Galen's λυκοπέρσιον was an Egyptian plant.

Lycopodium, name, in Tabernaemontanus, of *L. clavatum,* from λύκος, wolf, and πόδιον, little foot, trans. of a German name *Wolfsklauen.*

Lycopsis, λύκοψις, name of a plant, in Dioscorides, resembling ἄγχουσα, apparently from λύκος, wolf, and ὄψις, appearance.

Lycopūs, λύκος, wolf, and πούς, foot. Application doubtful.

lȳdius, a, um, of Lydia, a country in Asia Minor.

Lȳsimachia, λυσιμάχιον (and -χία), name of plant called after Lysimachos, king of Thrace. The word λυσίμαχος means ending strife, and the English name loose-strife is a translation of this word.

Lythrum, λύτρον, name of a plant in Dioscorides, called also λυσυμάχειος. Our spelling suggests the word λύθρον, gore.

macer, cra, crum, lean, meagre.

macro-, large- (μακρός originally meant long).

macrorrhīzus, a, um, having a large root (ῥίζα, root).

Macrotomia, μακρός, long, τόμος, segment, from the long calyx segments.

maculātus, a, um, spotted, speckled (p.part. of *maculo,* make spotted, *macula,* spot, mark).

Madia, the Chilean name of *M. sativa.*

Mahonia, called after Bernard M. Mahon, American horticulturist.

Maianthemum, *Māius,* the month of May, and ἄνθεμον, flower.

mājālis, e, used in botany to signify flowering during May (*Māius*).

mājor, us (comparative of *magnus*), greater.

Malachium, from μαλακός, soft.

Malaxis, μάλαξις, softening, because of the soft foliage.

Malcolmia, called after William Malcolm, eighteenth-century English horticulturist.

Mālus, the Latin name of the apple-tree.

Malva, plant-name in Pliny, who distinguishes two kinds, *malva sativa,* cultivated, and *m. silvestris,* wild mallow. See *silvestris* and *sativus.*

manicātus, a, um, furnished with long sleeves (*manicae*). The bases of the petioles of *Gunnera manicata* are invested in sleeve-like tufts of pinnately laciniate scales.

margarītāceus, a, um, pearl-like (*margarīta*, pearl).

Marianus, a, um, said to refer to Our Lady. According to legends, the white marks on the leaves of *Silybum Marianum* represent drops of milk spilt when feeding the Infant Jesus during the flight to Egypt.

marīnus, a, um, belonging to the sea, growing immersed in the sea (*mare*, sea).

Mariscus, *juncus mariscus,* name of a rush-like plant in Pliny.

maritimus, a, um, belonging to the sea, growing on the sea-coast (*mare*, sea).

Marrubium, name, in Pliny, of a plant, perhaps called after the city of the same name in Latium.

Martagon, ultimately from the Turkish name of a special form of turban adopted by Sultan Muhammed I (cf. the English name Turk's-cap Lily).

Marūta, Italian name of *Anthemis Cotula.*

masculus, a, um (male), vigorous, bold, manly, furnished with testicle-like tubers.

Mātrīcāria, medieval name, perhaps of *Matricaria Chamomilla,* from *mātrix, īcis,* womb, from its use in uterine disorders.

matritensis, e, of Madrid.

mātrōnālis, e, of or belonging to a married woman (*mātrōna*). Cf. Dame's Violet, other vernacular names, and the old Latin name *viola flos matronalis.*

Matthiola, called after Pierandrea Mattioli, famous Italian Renaissance botanist, author of *Commentarii in sex libros Pedanii Dioscoridis.*

maximus, a, um (superlative of *magnus*), largest, very large.

Mēcōnopsis, μήκων, ωνος, poppy, and ὄψις, appearance.

Mēdicāgo, word formed by adding the feminine suffix -*āgo* to *Mēdica* (μηδική), which was the name of *M. sativa,* a Persian or Median crop.

Mēdium, substantive, μήδιον, name of a plant in Dioscorides, perhaps *Campanula lingulata.*

medius, a, um, middle, intermediate (in size).

Melampȳrum, μελάμπῡρον, name in Theophrastus of a plant growing among wheat, from μέλας, black, and πῡρός, wheat. The seeds of MELAMPYRUM resemble black grains of wheat.

Melandrium, origin obscure. *Malundrum* was the name, in Pliny, of a plant, which was possibly *Melandrium album* (= *Silene alba*).

51

Melanion, μελάνιον, violet (μέλαν ἴον, black-violet; cf. *Leucojum* and *Nomimium*).

melano-, black (μέλᾱς, μέλᾰνος).

Meleāgris, the Guinea-fowl (*Numida meleagris* L.). *Fritillaria Meleagris* is so called from its chequered perianth.

Melica, name, in Cesalpino, of a kind of SORGHUM, perhaps connected with μελίνη, name, in Herodotus, of *Setaria italica*. In Lombardy *Sorghum vulgare* is still called *melga* and *melgone*.

Melilōtus, μελίλωτος, name in Theophrastus of a kind of clover, from μέλι, honey, and λωτός, name of various plants.

Melissa, μέλισσα, bee, nymph who kept bees. Medieval name of *Melissa officinalis* referring to its use in apiculture.

Melissophyllum, name of a plant in Pliny (bee-leaf).

melitensis, e, belonging to the Isle of Malta (*Melita*).

Melittis, from μέλιττα, Attic for μέλισσα, bee. See MELISSA.

Mentha (*Menta*), mint (μίντη). The plant is mentioned by Pliny and the myth associated with it is in Ovid.

Mēnyanthes, μήνάνθος, name, in Theophrastus, of *Nymphoides peltata*. Perhaps from μηνύω, disclose, and ἄνθος, flower, from the conspicuous flowers, or from μιννανθής, blooming for a short time.

Menziesia, called after Archibald Menzies (1754–1842), surgeon-botanist on Vancouver's famous expedition, from which he brought back numerous plants.

Mercuriālis, *herba mercuriālis,* name in Cato of a plant called after the god Mercury.

Mertensia, called after Franz Karl Mertens (1764–1831), botanical author.

-merus, a, um, having so many parts; as *tetramerus* (4-merous), having parts in fours (μέρος, part).

Mesēmbrianthemum, μεσημβρία (μέσος, ἡμέρα), mid-day, and ἄνθεμον, flower, from the flowers opening at noon. The generally accepted spelling with *y* for *i* is based on a derivation from μέσος, middle, ἔμβρυον, embryo, and ἄνθεμον, flower, meaning, according to Dillenius, a flower with an embryo in its middle.

Mespilus, name, in Pliny, of several trees; μεσπίλη σατάνειος of Theophrastus was probably *M. germanica*.

Metachlamydeae, fem.pl.adj. (sc. *plantae*), from μετά, here meaning beyond or over, and χλαμύς, ύδος, cloak (perianth), referring to the more advanced grade of elaboration of the perianth. See *Archichlamydeae*.

Mēum, μῆον, name, in Dioscorides, of a plant thought by some to be *Meum athamanticum*.

Mezereum, ultimately from the Persian name.

Mibora, equivocal word invented by Adanson.

micāceus, a, um, growing on mica (*mīca*, crumb).

Micranthēs, from μῑκρός, small, and ἄνθη, flower.

micro- (μῑκρός), small-.

Microcāla, μῑκρός, small, and κᾰλος, beautiful.

microglōchin, μῑκρός, small, and γλωχίν, point, from the bristle-like prolongation of the axis of the spike of *Carex microglochin.*

microptilon, μῑκρός, small, and πτίλον, anything like a feather or wing, referring to the narrow phyllary appendages.

mīlitāris, e, war-like, soldier-like (*mīles, itis,* soldier).

Milium, the Latin name of a millet.

Millefolium, classical name of several plants with finely cut leaves (*mille,* thousand, and *folium,* leaf). See *Myriophyllum.*

Millegrāna, *mille,* a thousand, and *grānum,* seed, fruit.

Mīmulus, diminutive of *mīmus,* mimic actor, from the mask-like corolla.

miniātus, a, um, coloured with, of the colour of, cinnabar (*minium*). The word is the p.part. of *minio,* to colour with cinnabar.

minimus, a, um (superlative of *parvus*), least, smallest, very small.

minor, us (comparative of *parvus*), smaller.

Minuartia, called after Juan Minuart of Barcelona (1693–1768), botanical author.

mītis, e, mild, mellow, bland, opposite of *ācer* and *acerbus.*

Moehringia, called after Paul Heinrich Moehring (1720–92), physician and naturalist at Jever in Oldenburg.

Moenchia, called after Konrad Moench (1744–1805), of Marburg, botanical author.

Molinia, called after G. B. Molina, writer on Chilean botany.

Mōlium, called after *Allium Moly* (μῶλυ, name of a magic herb, later applied to ALLIUM spp.).

mollis, e, soft.

Mollūgo, name of a plant in Pliny. The word is now used as the name of a genus of *Aizoaceae* (*mollis,* soft, and feminine suffix *-ūgo*).

molucellifolius, a, um, having leaves like those of *Molucella levis,* an eastern Mediterranean plant sometimes found naturalized in Britain.

monensis, e, pertaining to the Isle of Man or to Anglesey, both of which were called *Mona.*

Monermeae, called after the genus MONERMA (μόνος, single, and ἕρμα, prop, support, referring to the single glume).

Moneses, μόνος, single, and ἕσις, a sending out (ἵημι, send), referring to the single, large flower.

53

monīliformis, e, like a string of beads (*monīle*, necklace; Rich, p. 431).

mono- (before vowels **mon-**), μόνος, alone, single.

monogynus, a, um, μόνος, single, and γυνή (woman), carpel, style.

Monorchis, μόνος, single, and ὄρχις, testicle. At the time of flowering usually only one tuber is present.

Monotropa, μονότροπος, solitary, referring to the sequestered habitat in dark woods, or from μόνος, single, and τροπή, turn, direction, referring to the secund inflorescence.

monspeliensis, e, of Montpellier.

montānus, a, um, growing in mountainous places (*mons, tis*, mountain).

Montia, called after Giuseppe Monti, Professor of Botany at Bologna.

monticola, substantive, a dweller in mountains (*mons, tis*, mountain, and *colo*, inhabit).

Moricandia, called after Moïse Etienne Moricand (1780–1854), of Geneva, botanical author.

Mōrio, *mōrion* (μώριον), name of a plant which caused madness (μωρία).

Morsus-rānae, name first used by de l'Obel, from *morsus*, bite, and *rānae*, genitive of *rāna*, frog.

Moschatellīna, medieval name of *Adoxa Moschatellina*, fem. double diminutive of *moschātus*.

moschātus, a, um, smelling like musk (μόσχος, musk).

mūcrōnātus, a, um, ending in a short, straight point (*mucro, ōnis*, point, especially of a sword).

Mūgo, the Italian name of the tree in South Tirol.

Mulgēdium, word formed by Cassini from *mulgeo*, to milk, in analogy with LACTUCA (q.v.) from *lac*, milk. The two genera are scarcely distinct.

multi-, multus, a, um, many.

mūrālis, e, of, growing on, walls (*mūrus*, wall).

mūricātus, a, um, covered with short, hard tubercles (original meaning, shaped like the point on the end of the shell of the shell-fish *Mūrex* (Rich, p. 436)). See *purpurātus*.

mūrīnus, a, um, of mice, mouse-like (*mūs, mūris*, mouse). The combination *hordeum murinum* occurs in Pliny.

mūrōrum, of walls (gen.pl. of *mūrus*, wall).

Muscari, from the Arabic name of *M. moschatum*.

muscifer, a, um, fly-bearing, from *musca*, fly, and *fero*, bear.

muscipulus, a, um, badly formed word meaning fly-catching (*musca*, fly, and *capio*, catch). Flies adhere to the viscid stems of *Silene muscipula*. In classical Latin *muscipula* meant a mouse-trap (*mūs*, mouse, and *capio*).

muscōsus, a, um, moss-like (*muscus*, moss).

myagroidēs, resembling MYAGRUM.

Myagrum, μυάγρα, μύαγρον, mouse-trap, also a plant-name in Dioscorides (μῦς, mouse, and ἄγρα, hunting, the chase).

Mycelis, meaningless word coined by de l'Obel.

Myosōtis, μυοσωτίς, μυὸς ὦτα, name of 2 plants in Dioscorides, from μῦς, mouse, and οὖς, ὠτός, ear.

Myosōton, synonym, in Dioscorides, of *Myosotis.*

Myosūrus, mouse-tail (see *Myurus*), from the form of the fruit.

Myrīca, μυρίκα (μυρίκη), name of a species of TAMARIX.

Myriophyllum, μυριόφυλλον, name, in Dioscorides, probably of MYRIO-PHYLLUM sp., from μυρίος, countless, and φύλλον, leaf, referring to the finely divided leaves.

Myrrhis, μυρρίς, name, in Dioscorides and Theophrastus, of a plant thought by some to be *Myrrhis Odorata.*

myrsinītēs, intended for myrtle-like. The Greek word μυρσινίτης meant wine flavoured with myrtle (μυρσίνη, myrtle, and -ίτης; see -ītēs).

Myrtillus, said to be a diminutive of *myrtus*, myrtle.

Myrtus, myrtle (*Myrtus communis*).

Myūrus, mouse-tail (μῦς, μυός, mouse, and οὐρά, tail), from the form of the panicle.

Nāias, νāϊάς, water nymph (νάω, flow).

nāna, ae, female dwarf.

nānus, i, dwarf, also **a, um,** as modern adjective.

Nāpellus, diminutive of *nāpus*, turnip, from the form of the root-stock.

Nāpus, name of turnip in Pliny and Columella.

Narcissus, νάρκισσος, name of several plants probably belonging to the genus NARCISSUS, perhaps from νάρκη, torpor, referring to supposed narcotic properties.

Nardosmia, νάρδος, spikenard, and ὀσμή, odour, referring to the fragrant flowers.

Nardus, νάρδος, the name of *Nardostachys Jatamansi (Valerianaceae)*, the spikenard of the Bible, which the basal portions of *Nardus stricta* somewhat resemble (cf. *Spicant*).

Narthēcium, said by the author of the genus to be from νάρθηξ, rod, the stem being like an erect rod. In Pseudo-Dioscorides ναρθήκιον was a synonym of ἀσφόδελος, which was probably *Asphodeline liburnica*. The plant bears some resemblance to a miniature *Asphodeline liburnica* or FERULA (νάρθηξ). NARTHECIUM happens to be an anagram of ANTHERICUM, the name of a genus allied to ASPHODELUS and ASPHODELINE. ANTHERICUM is, however, a modern mis-spelling of the plant-name ἀνθέρικος.

55

Nasturtium, name of a pungent plant in Pliny, who wrote 'nomen accipit a narium tormento' (*nāsus*, nose, and *torqueo, tortum,* twist).

natans, swimming, floating (pres.part. of *nato*, swim).

Naumburgia, called after J. S. Naumburg (1768–99), Professor of Botany at Erfurt.

neāpolītānus, a, um, of Naples (Neāpolis, a name corresponding to our 'Newton', from *νέος*, new, and *πόλις*, city).

neglectus, a, um, neglected, overlooked, cf. **praetermissus.**

nemorālis, e, belonging to woods (*nemus, oris,* woodland).

nemorōsus, a, um (full of woods, shady), growing in woods (*nemus, oris,* wood).

nemorum, of woods (gen.pl. of *nemus, oris,* wood).

neocorymbōsus, a, um, new name for one of the *Hieracium corymbosum* segregates (*νέος*, new).

Neotinea, *νέος*, new, and TINEA, name of the genus, called after the Sicilian botanist Tineo.

Neottia, *νεοττιά* (Attic form of *νεοσσιά*), nest of young birds, from the appearance of the roots.

Nepeta, name of a plant in Pliny, etc., also the name of a city in Etruria.

Nephrōdium, *νεφρός*, in pl., kidneys, from the form of the indusium. Long *o* from compound *νεφρώδης = νεφροειδής*, kidney-like.

Nephrophyllum, *νεφρός*, the kidneys, and *φύλλον*, leaf, from the reniform leaves.

nericius, a, um, of Närke, province in Sweden.

Neslia, called after Nesles, an eighteenth- and nineteenth-century French botanist.

nessensis, e, of Loch Ness, Inverness-shire.

nīcaeensis, e, Nicene, of Nicaea, a city in Bithynia.

Nidus-avis, *nidus*, nest, and *avis*, gen. *avis*, bird. See NEOTTIA.

Nigella, nigellus, a, um, somewhat black, diminutive of *niger*, referring to the black seeds of *N. sativa*, which are used as a condiment.

niger, gra, grum, black.

nigrescens = *nigricans*.

nigricans, blackening, especially when dry.

Nintooa, from *nin too*, the Japanese name of *Lonicera japonica*.

Nissolia, called after Guillaume Nissole, seventeenth-century botanist at Montpellier.

nitens, shining (pres.part. of *nīteo*, shine).

nivālis, e, growing in or near snow (*nix, nivis,* snow).

niveus, a, um, snow-white (*nix, nivis,* snow).

56

nōbilis, e, famous, noted, celebrated.

noctiflōrus, a, um, flowering at night (*nox, ctis,* night, and *flōs, ōris,* flower).

nōdōsus, a, um, knotty, referring to tubers or swollen nodes (*nōdus,* knot).

nōli-[mē-]tangere, do not touch [me]. See IMPATIENS.

Nomimium, from νόμιμος, customary, and ἴον, violet, i.e. the usual type of violet.

nōn-scriptus, a, um, not written upon. See *Ajācis* and HYACINTHUS.

nootkatensis, e, native of Nootka Sound, on west coast of British Columbia. The *t* is inserted for the sake of euphony.

Nothoscordum, νόθος, spurious, and σκόρδον, garlic.

nōto-, νῶτον or νῶτος, the back, any wide surface, surface of cotyledons.

Novi-belgii, of New York, which was from 1614–64 a Dutch settlement, New Netherlands (*Novum Belgium*). The area occupied by the ancient *Belgae* included modern Belgium and Holland.

nūdi-, nūdus, a, um, naked-.

nūdicaulis, e, having a leafless stem (*nūdus,* naked, and *caulis,* stem).

nummulārius, a, um, in botanical Latin means having circular (coin-like) leaves (*nummus,* coin).

Nuphar, from the Persian *nūfar, naufar,* water-lily.

nūtans, nodding, drooping (*nūto,* nod).

Nymphaea, νυμφαία, name of a water-plant in Theophrastus (νύμφη, (water) nymph).

Nymphoīdēs, NYMPHAEA and *-oīdēs,* resembling.

Obiōnē, probably from the Siberian river Ob or Obi and fem. patronymic suffix *-ώνη,* 'daughter of the Obi'.

obscūrus, a, um, dark, dusky, referring, in *Centaurea obscura,* to the black involucres. Also means obscure, of uncertain identity or affinities.

obvallāris, e, surrounded by a rampart or wall. The indented corona of *Narcissus obvallaris* resembles a battlement.

occidentālis, e, western.

occitānicus, a, um, of the province Occitania, now called Languedoc.

ōchroleucus, a, um, ὠχρόλευκος, yellowish white (ὠχρός, pale yellow, and λευκός, white).

-ōdēs, -ώδης, suffix denoting connexion with or resemblance to.

Odontītis (-ēs), name, in Pliny, of a plant good for toothache (ὀδούς, ὀντος, tooth, and suffix *-ῖτις,* fem. of *-ίτης,* meaning connected with).

57

odōrātus, a, um, sweet-smelling, fragrant (*odōro,* scent sweetly).

Oederi, called after Georg Christian Oeder (1728–91), author of the first three volumes of *Flora Danica.*

oedocarpus, a, um, οἰδέω, become swollen, and καρπός, fruit.

Oenanthē, οἰνάνθη, name of a plant smelling like the vine, from οἶνος, wine, and ἄνθη, flower.

Oenothēra, οἰνοθήρας, ὀνοθήρας, names, in Greek authors, probably of *Nerium Oleander,* the latter perhaps the more correct form, meaning ass-catcher. Names, in various languages, of the Oleander signify ass poison. See *Guide to Cambridge University Botanic Garden,* 1st ed., p. 100. Much has been written about the derivation of this word. See also *Onagraceae.*

officinālis, e, kept at the druggist's 'shop' (*officīna*), i.e. used medicinally. The English word officinal now means included in the *British Pharmacopoeia.*

-(o)idēs, -like (-(o)ειδής, εἶδος, form, shape, with joining vowel o).

Oleāceae, called after the genus OLEA. *O. europaea* is the Olive Tree.

olerāceus, a, um, eaten as a vegetable ((*h*)*olus, eris,* vegetables, greens).

olidus, a, um, smelling, stinking.

olitōrius, a, um, used as a vegetable (*olitor,* kitchen gardener, *olus, eris,* potherb; these words should be spelt with an initial *h*).

Olusātrum, name, in Pliny and Columella, of *Smyrnium Olusatrum,* so called because of its conspicuous black fruits ((*h*)*olus,* potherb, and *āter,* jet black).

Omphalōdēs, ὀμφαλώδης, navel-like, referring to the form of the nutlets.

Omphalospora, from ὀμφαλός, navel, and σπορά, seed. The seeds in this section have a deep depression (see *Pocilla*).

Onagrāceae, fem.pl.adj. (sc. *plantae*) formed from ONAGRA, old name for OENOTHERA q.v. (ὄναγρος means wild ass, from ὄνος, ass, and ἄγριος, wild). In Dioscorides ὀνάγρα is a synonym of ὀνοθήρας.

Onobrȳchis, ὀνοβρῡχίς (ῠ in Liddell and Scott), name, in Dioscorides and Galen, of a leguminous plant thought to be *O. Caput-galli,* presumably from ὄνος, ass, and βρύχω, eat greedily, hence called *palmes asini* by Pliny.

Onōnis, ὄνωνις (ἄνωνις), name of a plant in Dioscorides.

Onopordon, ὀνόπορδον, name of a kind of thistle, from ὄνος, ass, and πορδή, noisy expulsion of wind, so called because of its effect when eaten by donkeys.

Onopteris, name used by Tabernaemontanus, from ὄνος, ass, and πτερίς, fern.

Ophioglōssum, ὄφις, snake, and γλῶσσα, tongue, from the appearance of the spike.

58

Ophioscordon, ὀφιοσκόρδον, name of a garlic in Dioscorides, probably from ὄφις, snake, and σκόρδον, garlic, referring to the twisted stem.

Ophrys, name, in Pliny, of a plant with two leaves, perhaps LISTERA or PLATANTHERA sp. ὀφρύς means eyebrow.

Opōrina, from ὀπωρινός, used as a translation of *autumnālis*. The word actually refers to late summer (see lexicons).

Opulus, name of a tree, probably a species of ACER.

ōrārius, a, um, belonging to the shore (*ōra*, shore).

orcadensis, e, of the Orkney Islands (*Orcades*).

Orchis, ὄρχις, name, in Dioscorides, of a plant with twin tubers like testicles: ὄρχις means testicle, and is a masculine word. The genitive ends in εος or εως, but a *d*-stem is used, for euphony, in compounds, as *Orchidaceae*.

Orēadea, called after *Hieracium oreades* (ὀρειάς, άδος, mountain-nymph).

Oreopteris, ὄρος εος, mountain, hill, and πτερίς, fern.

orientālis, e, eastern.

Orīganum, ὀρίγανον, name, in Theophrastus, etc., of an aromatic herb.

Ormenis, meaning obscure; ὄρμενος means a shoot, sprout, stem, or stalk.

Ornīthogalum, ὀρνῑθόγαλον, name of a plant in Dioscorides, from ὄρνις, ὄρνῑθος, bird, and γάλα, milk. The flowers of *O. nutans* resemble bird-droppings.

ornīthopodus, a, um, like birds' claws (see below).

Ornīthopūs, ὄρνις, ὄρνῑθος, bird, and πούς, foot. The fruits resemble birds' claws.

Orobanche, ὄροβος (Latin *ervum*), name of a leguminous plant, and ἄγχω, strangle. *O. crenata* is a pest on bean-fields in the Mediterranean region.

Orobus, ὄροβος, name, in Theophrastus, of a leguminous plant.

Orontium, ὀρόντιον, name of a plant in Galen.

ortho-, ὀρθός, straight, erect.

Ortholobum, ὀρθός, straight, and λοβός, lobe, here referring to the cotyledons.

oryzoïdēs, like ORYZA. Our species of LEERSIA is closely allied to *O. sativa*, the rice-plant (ὄρυζα, rice).

Osmunda, said to be derived from an epithet of the god Thor.

Osproleōn, ὀσπρολέων, name of a weed, probably *Orobanche crenata*, injurious to leguminous crops, from ὄσπρος (ὄσπριον), beans, and λέων, lion. See OROBANCHE.

ossifragus, a, um, bone-breaking, from *os*, bone, and *frango*, break. *Narthecium ossifragum* was said to make the bones of cattle feeding on it brittle. It grows on lime-free soils.

Ostruthium, name, in herbals, of *Peucedanum Ostruthium*. Various derivations are suggested. See Hegi, v, 2, p. 1363, n. 10.

Ōtanthus, οὖς, ὠτός, ear, and ἄνθος, flower, from the two spurs of the corolla. Cf. DIOTIS.

Ōtītēs, ὠτίτης means to do with ears. The lower leaves of *Silene Otites* resemble ear-picks.

Otrubae, of Josef Otruba (b. 1889) of Moravia, interested in CAREX.

ovīnus, a, um, to do with sheep. *Festuca ovina* is an excellent pasture grass (*ovis*, sheep).

Oxalis, ὀξαλίς, name, in Nicander, of *Rumex Acetosa* (ὀξύς, sharp, acid).

oxy-, ὀξύς, sharp, acute, acid.

Oxyacantha, ὀξυάκανθα, name in Theophrastus of *Pyracantha coccinea*, from ὀξύς, sharp, and ἄκανθα, thorn.

Oxycoccos, ὀξύς, acid, and κόκκος, berry.

oxypterus, a, um, having sharp wings (ὀξύς, sharp, and πτερόν, wing).

Oxyria, from ὀξύς, acid, referring to the taste.

Oxytropis, ὀξύς, sharp, τρόπις, keel; referring to the mucronate carina.

pachy-, stout-, thick- (παχύς).

Padus, πάδος, name in Theophrastus of *Prunus mahaleb*.

Paeōnia, name of a plant in Theophrastus, called after Παιάν (Ionic Παιών), physician of the gods.

pāgānus, a, um, belonging to the country (*pāgus*, country district).

paleāceus, a, um, furnished with chaffy scales (*palea*, chaff).

pallens, pres.part. of *palleo*, be pale.

pallescens, turning pale (pres.part. of *pallesco*, become pale).

palūdōsus, a, um = *paluster*.

paluster, tris, tre, growing in swampy places (*palus, ūdis*, swamp, bog). A masculine in -*tris* is often used in botanical Latin.

pampinōsus, a, um, leafy (*pampinus*, vine-foliage).

pāniceus, a, um, resembling the grain of the millet, *Panicum miliaceum*, as the ripe perigynia of *Carex panicea*.

pāniculātus, a, um, having a panicle, a term loosely used for a compound inflorescence considerably longer than broad (*pānicula*, tuft, thatch).

Pānicum, the Latin name of *Setaria italica*.

pannonicus, a, um, of Pannonia, a country which lay N. of Dalmatia.

panormitānus, a, um, of Palermo in Sicily (Πάνορμος was the name of Palermo and also of other seaport towns. The word is derived from πᾶν, all, and ὅρμος, anchorage, meaning always fit for anchorage).

Papāver, the Latin name of various members of the family, especially *Papaver somniferum*.

Pāpiliōnātae, fem.pl.adj. (sc. *plantae*) from *pāpilio, ōnis*, butterfly, referring to the form of the corolla.

paradoxus, a, um, 'paradoxical' (παρά, in sense of wrong, irregular, and δόξα, opinion). The word, when applied to plants, refers to some peculiarity.

Paralias, παράλιος, by the sea, term applied to maritime plants (παραλία, beach, παρά, beside, along, and ἅλς, salt).

Parapholis, παρά, beside, and φολίς, scale. The two glumes, when both are present, are situated side by side external to the rachis.

Pardalianches, παρδαλιαγχές, name, in Aristotle, of a plant used for poisoning beasts of prey, from πάρδαλις, leopard, and ἄγχω, strangle.

pardalinus, a, um, spotted like a leopard (πάρδαλις). The *i* may be lengthened if the word is considered Latin.

Parentucellia, called after Th. Parentucelli, an important figure in the Revival of Learning. He was born in Liguria, became Pope Nicholas V (1447–55), and founded the Vatican Library and the Botanic Garden at Rome.

Parietāria, name of a plant growing on walls, from *parietārius, a, um,* of walls, from *pariēs, etis,* wall (cf. 'parietal' placentation).

Paris, from *pār, paris,* equal, from the isomerous leaves and sporophylls.

Parnassia, called *grāmen Parnassi* (grass of Parnassus) by de l'Obel.

Paronychia, παρά, beside, and ὄνυξ, υχος, finger-nail. The plant was used for whitlows.

Parthenium, παρθένιον, name of a plant in Theophrastus, from παρθένος, virgin.

Parthenocissus, from παρθένος, virgin, and κισσός, ivy; formed on analogy with 'virginian creeper'.

parvulus, a, um, very small.

parvus, a, um, small.

Pastināca, name of the carrot, later used for the parsnip, probably from *pastino,* dig and trench the ground.

pastōrālis, e, belonging to shepherds, growing in pastures (*pastor, ōris,* shepherd).

patellāris, e, resembling a *patella* or small dish, referring to the corona of *Narcissus majalis* var. *patellaris.*

patens, spreading (pres.part. of *pateo,* be open).

patientia, from the French name *patience,* which perhaps arose from *lapathum* (q.v.), with initial *la-* regarded as article (see appendix).

patulus, a, um, spreading.

paucus, a, um, few.

pauperculus, a, um, poor.

61

Pecten-Veneris, comb of Venus, name used by Pliny, perhaps for SCANDIX sp.

pectinātus, a, um, comb-like (p.part. of *pectino*, comb).

Pedīculāris, *pedīculāris, e,* of, or belonging to lice. *Herba pedīculāris* was the name of a plant in Columella (*pedīculus*, louse).

pellūcidus, a, um, transparent, pellucid. *Hieracium pellucidum* has pellucid dots on the leaves (*perlūcidus, per,* through, and *lūceo,* shine).

peltātus, a, um, having the petiole attached to the lower surface of the lamina (original meaning, armed with a pelta or shield; see Rich, p. 487).

pendulīnus, a, um (for *pendulus, a, um*), hanging down.

pendulus, a, um, hanging down.

pēnicillātus, a, um, furnished with a tuft like the hairs on a painter's brush (*pēnicillum*, painter's brush (Rich, p. 488)).

pent-, penta-, 5- (πέντε, 5).

Pentaglōttis, πέντε, 5, and γλῶττα, Attic for γλῶσσα, tongue, referring to the 5 lingulate scales at the throat of the corolla.

Peplis, πεπλίς, name of *Euphorbia peplis* in Dioscorides.

peploīdēs, resembling PEPLUS. See *-oides.*

Peplus, πέπλος, name of *Euphorbia peplus* in Dioscorides.

Pepo, πέπων, cooked by the sun. σίκυος πέπων, name of a gourd which was eaten when ripe.

peregrīnus, a, um, foreign, exotic (*pereger*, abroad, away from home, from *per,* through, and *ager,* field, land).

perfoliātus, a, um, having the leaf completely embracing the stem, so that the stem apparently passes through the leaf (*per,* through, and *folium,* leaf).

Periclymenum, περικλύμενον, name, in Dioscorides, of a twining plant, thought to be *Lonicera etrusca.*

permixtus, a, um, confused (with other species), p.part. of *permisceo,* intermingle, throw into confusion.

perpropinquus, a, um, very near, closely allied.

perpusillus, a, um, very small.

Persicāria, name, in Rufinus, of *Polygonum Hydropiper,* meaning peach-leaved. See *persicifolius.*

persicifolius, a, um, with leaves like those of the peach-tree, which Pliny called *persica mālus.*

persicus, a, um, Persian.

Petasītēs, name in Dioscorides of *P. hybridus,* so called from the large leaves resembling the broad-rimmed hat called πέτασος. See Rich, p. 497.

petecticālis, e, spotted (modern Latin *petechia*, spot, from Italian *petecchia*, spot in fever, etc.).

petraeus, a, um, growing among rocks (πετραῖος = *saxatilis*; πέτρα, rock).

Petroselīnum, πετροσέλινον, name of parsley in Dioscorides, Galen, etc., from πέτρος, stone, and σέλινον (*see* SELINUM), perhaps from growing in stony places, or being good for stone in the bladder, or both.

Peucedanum, πευκέδανον, name of a plant in Theophrastus (see app.).

phaeo-, brownish- (φαιός).

phaeus, a, um, latinized φαιός, ά, όν, dusky, dun, brown.

Phalāris, φαλᾱρίς in Dioscorides, a kind of grass. Perhaps connected with *phalerae*, τὰ φάλαρα, smooth, shining, boss-like ornaments for men and horses. See Rich, p. 499.

Phēgopteris, word invented by Linnaeus on the analogy of DRYOPTERIS q.v. In Greek φηγός was a species of oak, and not the beech, though the word is cognate with *fāgus*.

Phellandrium, *phellandrion,* name, in Pliny, of a plant with ivy-like leaves.

Philadelphus, φιλάδελφος, the name, in Athenaeus, of a sweet-scented shrub, perhaps called after Ptolemaeus Philadelphus, king of Egypt. The word means loving one's brother or sister, but it was also a Grecian and Roman surname.

Philonotis, moisture-loving (φίλος, friend, and νοτίς, moisture).

-philus, a, um, -loving (φίλος, friend).

Phleum, φλέως, name in Greek authors, said to be of *Erianthus Ravennae,* perhaps applied to our genus because of the crowded inflorescence (φλέων, teeming, pres.part. of φλέω).

Phlomis, φλομίς, VERBASCUM spp. (see appendix).

phoenīceus, a, um, red-purple, from Φοινίκη, Phoenicia, a country fence, and -ίτης, connected with) (see appendix).

phoenīcolasius, a, um, φοῖνιξ, ῑκος, a crimson dye, thought to have been discovered by the Phoenicians, and λάσιος, hairy, shaggy, referring to the copious red hairs on the shoots.

Pholiūrus, badly formed word from φολίς, ίδος, scale, and οὐρά, tail, referring to the tail-like spikes densely beset with scale-like glumes.

Phragmītēs, κάλαμος φραγμίτης, a kind of reed growing in hedges (Dioscorides). The Greek adjective means of or for a fence (φράγμα, fence, and -ίτης, connected with).

phylicifolius, a, um, having leaves like those of PHYLICA, an African genus of *Rhamnaceae* (φυλίκη, name of a plant in Theophrastus).

63

Phyllītis, φυλλῖτις, name, in Dioscorides, of *Phyllitis Scolopendrium.*

Phyllodocē, name of a sea nymph.

Phyllodolon, φύλλον, leaf, and δόλος, cunning contrivance for catching, net, referring to the sheathing leaves.

-phyllus, a, um, -leaved (φύλλον, leaf).

Phȳsospermum, φῦσα, bellows, and σπέρμα, seed, referring to the inflated fruits.

Phyteuma, the word φύτευμα, originally meaning 'that which is planted', was used by Dioscorides as the name of a definite plant, probably *Reseda Phyteuma* (φυτεύω, plant, and -μα, neuter suffix, denoting result of action).

Picea, name of a pitch-yielding tree (*pix, icis,* pitch).

Picris, πικρίς, name, in Theophrastus, of a bitter herb (πικρός, bitter).

Pilosella, name, in Rufinus, of *Hieracium Pilosella* (fem.dimin. of *pilōsus*).

pilōsissimus, a, um, superlative of *pilōsus.*

pilōsus, a, um, hairy, in botanical Latin usually meaning with soft and distinct hairs (*pilus,* hair).

Pilulāria, *pilula,* diminutive of *pila,* ball, referring to the sporocarps.

pilulifer, a, um, bearing little balls, e.g. the globular carpellary inflorescences of *Urtica pilulifera, Carex pilulifera* (*pilula,* little ball, pill, and *fero,* bear).

Pimpinella, name first used by Matthaeus Sylvaticus. Various fanciful derivations have been suggested.

Pinardia, called after Pinard, superintendent of the Botanic Garden at Rouen.

Pīnaster, wild pine, from *pīnus,* and *-aster,* suffix meaning wild, equivalent to the adjective *silvestris.* According to Pliny, *pīnaster* and *pīnus silvestris* were one and the same.

Pinguicula, name first used by Gesner, fem.dimin. (sc. *planta*) of *pinguis,* fat, from the appearance of the leaves.

Pīnus, the Latin name of *Pinus Pinea* and other species.

piperītus, a, um, pepper-like (πεπερίζω, taste like pepper).

Pirola, diminutive of *pirus,* pear-tree, from the form of the leaves.

Pirus, the Latin name of the pear-tree.

Pīsum, *pīsum,* pea.

plantāgineus, a, um, with leaves like those of PLANTAGO.

Plantāgo, the Latin name of *P. major,* from *planta,* sole of the foot, referring to the broad, flat leaves pressed against the ground, and feminine termination *-āgo.*

Platanthēra, πλατύς, broad, and ἀνθηρά (in botany), anther, referring to the diverging anther-thecae of *P. chlorantha.*

platy-, broad- (πλατύς, broad).

platyphyllos, broad-leaved, πλατύς, broad, and φύλλον, leaf.

Plectolobum, πλεκτός, twisted, and λοβός, lobe (cotyledon).

Plēthosphace, apparently from πλῆθος, greater number, main body, crowd or mass, and σφάκος, SALVIA sp.

pleuro-, πλευρά, side, edge, rib.

plicātus, a, um, folded (p.part. of *plico*, fold).

plūmārius, a, um, embroidered with feathers (*plūma*, small soft feather).

Plumbagināceae, from the genus PLUMBĀGO. *Plumbago* was the name, in Pliny, of a plant also called μολύβδαινα. *Plumbum* and μόλυβδος both mean lead. The reference is perhaps to the colour of the flowers. The feminine suffix -*āgo* is common in plant-names.

Pneumonanthē, πνεύμων, ονος, lung, and ἄνθη, flower, perhaps because the spotted flower suggests the appearance of lung. See PULMONARIA.

Poa, πόα, grass, especially as fodder.

podagrārius, a, um, good for gout (*podagra*, gout in the feet).

poēticus, a, um, of poets (from Homer onwards).

Polemōnium, *polemōnia*, name of a plant in Pliny, probably called after King Polemon of Pontus.

polifolius, a, um, having grey leaves like those of *Teucrium Polium*, πόλιον (πολιός, grey).

polītus, a, um, polished, refined (p.part. of *polio*, polish).

poly-, πολύς, many, but the element sometimes has the force of separate, as in *Polypetalae*.

Polycarpon, πολύκαρπος, ον, rich in fruit, the neuter form was used as a plant-name by Hippocrates (πολύς, many, καρπός, fruit).

polycerātus, a, um, πολύς, many, and κέρας, ᾱτος, horn, referring to the numerous, curved pods.

Polychondreae, πολύς, many, and χόνδρος, granule. The pollinia are made up of numerous packets of pollen grains.

polycladus, a, um, having many shoots (πολύς, many, and κλάδος, shoot).

polyedrus, a, um, many-sided, polyhedric, referring to the fruits of *Sparganium polyedrum* (πολύς, many, and ἕδρα, base, side of solid figure).

Polygala, name of plant in Pliny, πολύγαλον in Dioscorides, from πολύς, much, and γάλα, milk, so called because it was supposed to increase the secretion of milk.

Polygonātum, πολυγόνατον, name of *P. multiflorum*, from πολύς, many, and γόνυ, ατος, knee (the rhizome is geniculate). Note the short *a*.

Polygonum, πολύγονον, name of a herb in Dioscorides, perhaps from πολύς, much, and γόνος, seed, progeny, from the numerous fruits. Bauhin used the word, in this sense, as a synonym of *millegrāna* q.v. See also *aviculāris*. The second element may be γόνυ, knee, referring to the swollen nodes.

polymorphus, a, um, multiform, of various shapes, from πολύς, many, and μορφή, shape, form.

Polypodium, πολύς, many, and πόδιον, diminutive of πούς, foot, of doubtful application.

Polypōgon, πολύς, many, and πώγων, beard, from the numerous, crowded awns.

polystachyus, a, um, πολύς, many, and στάχυς, originally ear of corn, used for various condensed paniculate inflorescences.

Polystichum, πολύς, many, and στίχος, row, referring to the multiseriate sori.

ponticus, a, um, belonging to the region of the Black Sea (*Pontus*).

Pōpulus, poplar.

Porphyrion, from πορφύρεος, purple, dark red, from the colour of the flowers.

porrifolius, a, um, having leaves like the leek (*porrum*).

porrigens, spreading (pres.part. of *porrigo*, spread).

Porrum, *porrum* was the Latin name of several species of ALLIUM. Cognate with πράσον (ă), which enters into the trivial names of several of our ALLIUM spp.

portensis, e, of Oporto (Porto).

Portula, abbreviated form of PORTULACA.

Portulāca, name, in Latin authors and in Italian, of *Portulaca oleracea*.

Potamogētōn, name of *P. natans* in Dioscorides, from ποταμός, river, and γείτων, neighbour.

Potentilla, dimin.subst. formed from *potens*, powerful, because of supposed medicinal virtues.

Potērium, ποτήριον, drinking cup; ποτίρριον was the name of a plant in Dioscorides, thought to be *Astragalus Poterium*.

praealtus, a, um, very high, used in botanical Latin for very tall.

praecox, ocis, early (flowering) (*praecoquo*, ripen).

praelongus, a, um, very long.

praeruptōrum, of steep or rugged places (gen. of pl.subst. *praerupta*).

praetermissus, a, um, overlooked, not previously described (p.part. of *praetermitto*, omit).

prātensis, e, growing in meadows (*prātum*, meadow).

pratericola, origin obscure, perhaps meant for *prāticola* or *prātincola*, substantive, from *prātum*, meadow, and *incola*, inhabitant.

Prēnanthēs, πρηνής (*prōnus*), facing downwards, and ἄνθη, flower, from the nodding young heads.

Prīmula, *primula vēris* (the first of the spring), old designation for early-flowering herbs. *Prīmula* is feminine of adj. *prīmulus*, diminutive of *prīmus*, first.

prōcērus, a, um, tall.

prōcumbens, lying along the ground (pres.part. of *prōcumbo,* extend, spread).

prōlifer, ra, rum, proliferous, i.e. reproducing by buds (*prōles,* offspring, and *fero,* bear). *Dianthus prolifer* is so called from the heads producing numerous flowers.

prōnus, a, um, leaning forward, lying face downwards.

propinquus, a, um, near, allied to.

Prunella. See BRUNELLA.

Prūnophora, plum-bearing, from *prūnum,* plum, and -φόρος, bearing, φέρω, bear.

Prūnus, *prūnus,* plum-tree.

Pseudacorus, ψευδο-, false, and ACORUS q.v.

Pseudo-Narcissus, ψευδο-, false, and NARCISSUS. *N. poëticus* was considered the only 'true' NARCISSUS.

Pseudo-Platanus, ψευδο-, false, and PLATANUS.

Pseudotsuga, ψευδο-, false, and TSUGA, name of a genus of *Pinaceae.*

Psyllium, ψύλλιον, name, in Greek authors, of *Plantago Psyllium.* The seeds of this species resemble fleas (ψύλλα, flea).

Ptarmica, πταρμική, name in Dioscorides of a plant with small camomile-like heads and leaves like those of the olive. It caused sneezing (πταίρω, sneeze).

Pteridium, πτερίδιον, diminutive of πτερίς, fern. The word πτερίδιος meant made of feathers, winged.

Pteridophy̆ta, neut.pl.subst., from πτερίς, ίδος, fern, and φῡτόν, τό, plant. Note the short *y.*

-pterus, a, um, -winged (πτερόν, wing).

pūbescens, covered with soft hairs.

Puccinellia, called after Benedetto Puccinelli, nineteenth-century Professor of Botany at Lucca.

Puelii, of T. Puel, nineteenth-century Parisian doctor, who did work on the French and Syrian floras.

pulchellus, a, um, pretty (diminutive of *pulcher,* beautiful).

pulcher, chra, chrum, beautiful.

Pūlegium, Latin name of a plant, perhaps used to dispel fleas (*pūlex,* flea).

Pūlicāria, name of a plant obnoxious to fleas (*pūlex, icis,* flea).

pūlicāris, e, to do with fleas (*pūlex, icis,* flea), from the appearance of the fruits of *Carex pulicaris.*

Pulmōnāria, the spotted leaves of *P. officinalis* (*Herba Pulmonariae maculosae*) were thought to be a signature of the lungs (*pulmōnes*). Hence the plant was used in pulmonary disorders.

Pulsātilla, name used by Brunfels, diminutive of *pulsāta*, beaten, driven about, according to Linnaeus from the beating of the flowers by the wind ('pulsatione floris vento').

pulverulentus, a, um, dusty, mealy (*pulvis, eris,* dust).

pūmilus, a, um, dwarfish, diminutive.

punctātus, a, um, marked with dots or small depressions (*punctum,* point, dot).

pungens, pricking (*pungo,* prick).

purpurātus, a, um, clad in *purpura,* i.e. the purple dye (Tyrian purple), obtained from the anal glands of the shell fish *Murex.*

purpureus, a, um, applied to various shades of red.

pusillus, a, um, very little, insignificant.

pycno-, πυκνός, dense, compact.

pycnocephalus, a, um, having heads arranged in dense clusters, πυκνός, dense, and κεφαλή, head.

pycnotrichus, a, um, having hairs arranged in dense tufts, πυκνός, dense, and θρίξ, τριχός, hair.

pygmaeus, a, um, pigmy, dwarf (ultimately from πυγμή, distance from elbow to knuckles).

Pyr-. See *Pir-.*

pȳramidālis, e, shaped like a pyramid.

quadri-, four-.

quadriradiātus, a, um, having four ray-florets.

quarternellus, a, um, probably mis-spelt diminutive of *quaternārius,* consisting of four each, referring to the 4-merous flowers.

Quercus, Latin name, probably, of *Q. Robur.*

-quetrus, a, um, with acute angles (*triquetrus,* triangular).

quinquevulnerus, a, um, *quinque,* 5, and *vulnus, eris,* wound. Each petal has a red blotch.

rādīcans, rooting, term usually applied to stems (pres.part. of *rādīco(r),* to strike root).

rādīcātus, a, um, rooted, having a large root (p.part. of *rādīco(r)*).

Radiola, name of a plant in the famous herbal called *Herbarium Apuleii Platonici,* perhaps a diminutive of *radius,* ray, from the radiating branches.

rādula, a scraping iron. *Rubus radula* is so called from the copious bristles on the shoots.

Ramischia, called after F. X. Ramisch, Professor at Prague (see app.).

rāmōsus, a, um, branched (*rāmus,* branch).

Ranunculus, diminutive of *rāna*, frog, used by Pliny as a synonym of *batrachion*. See *Batrachium*.

Rāpa (rāpum), name of turnip in Columella and Varro.

rāpāceus, a, um, turnip-like (*rāpum*, turnip).

Raphanus, word used for the radish in Pliny and Columella; ῥάφανος, meant cabbage in Attic Greece, radish elsewhere.

Rāpistrum, name, in Columella, meaning wild turnip, from *rāpum*, turnip, and suffix *-aster*, q.v.

Rāpum-Genistae, turnip of GENISTA. *Orobanche Rapum-Genistae* has a tuberously thickened stem-base and is parasitic on shrubby *Leguminosae*.

Rāpunculus, word first used by Bock as RAPUNCULUM, diminutive of *rāpum*, turnip, from the tuberous root.

rĕclīnātus, a, um, leaning back, reclining (see appendix).

redivīvus, a, um, that lives again, renewed. Perhaps referring to the fact that *Lunaria rediviva* is perennial.

rēgālis, e, kingly, royal, referring to the dignified appearance of *Osmunda regalis* (*rex, rēgis,* king).

rēgius, a, um, royal, princely, splendid (*rex, rēgis,* king).

remōtus, a, um, distant, as, for example, the spikes of *Carex remota* (p.part. of *removeo*).

repandus, a, um, with slightly wavy margin (classical meaning, bent backwards).

rēpens, creeping (*repo*, creep), usually applied to creeping stems which root at the nodes.

reptans = *repens* (*repto*, frequentative of *repo*).

Resēda, name in Pliny of a plant, possibly *Reseda alba*. The word is presumably derived from *resēdo*, heal.

resupīnātus, a, um, lying with face upwards, upside down (p.part. of *resupīno*).

rēticulātus, a, um, with a conspicuous network of veins (*rēte*, net).

Rhamnus, ῥάμνος, name of several prickly shrubs in various Greek authors. The Greek word is feminine: *rhamnos*, in Pliny, is masculine.

Rhāponticum, the ῥᾶ (see RHEUM), growing in the region of the Black Sea (see *ponticus, a, um*). Our word rhubarb is from *rhābarbarum*, meaning foreign (*barbarus, a, um*) ῥᾶ.

Rhēum, ῥῆον in Galen = ῥᾶ in Dioscorides, from a Persian name of RHEUM spp.

Rhīnanthus, ῥίς, ῥῑνός, nose, and ἄνθος, flower, referring to the trunk-like upper corolla lip of the Mediterranean species *R. elephas* L., now placed in the genus RHYNCHOCORYS.

-rhīzus, -a, -um, -rooted (ῥίζα, root).

69

Rhodiola, the root-stock, which smells like roses, was formerly called *Radix Rhodiae,* and *Rhodiola* is doubtless a badly formed diminutive of *Rhodiae* (ῥόδον, a rose).

Rhododendron, ῥοδόδενδρον, name in Latin and Greek authors, thought to belong to *Nerium Oleander* (ῥόδον, rose, and δένδρον, tree).

Rhoeadāles. See *Rhoeas.*

Rhoeas, μήκων ῥοιάς, the name of *Papaver Rhoeas* from the colour of the flower resembling that of *Punica Granatum* (ῥοιά).

Rhopalostachya, ῥόπαλον, club, cudgel (Rich, p. 172), and *stachyon,* q.v. The sporangia are borne in cones, which form club-like terminations to the branches. Cf. the English name Club-moss.

Rhynchosināpis, ῥύγχος, beak, and SINAPIS, from the conspicuously beaked fruits.

Rhynchospora, ῥύγχος, beak, and σπορά, seed (here meaning fruit), from the beaked nuts.

Ribes, from *rībās,* Arabic name of *Rheum ribes,* with acid juice.

rīmōsus, a, um, originally, full of cracks or fissures. *Valerianella rimosa* has a deep furrow in its fruit (*rīma,* cleft, furrow).

rīpārius, a, um, growing by rivers and streams (*rīpa,* river-bank).

rīvālis, e, belonging to rivers and brooks (*rīvus,* brook).

rīvulāris, e, growing beside brooks (*rīvulus,* diminutive of *rīvus,* brook).

Robertiānus, a, um, *Geranium Robertianum* was formerly *Herba Roberti.* It is not certain to which medieval Robertus or Ruprecht the name refers. *Thelypteris Robertiana* is so called because it smells like *Geranium Robertianum.*

Robinia, called after Jean Robin (1550–1629), of Paris, gardener to Henri IV and Louis XIII, who introduced *Robinia Pseudacacia* from Virginia in 1600. His son, Vespasian Robin, subsequently planted the tree extensively.

Roegneria, called after Rögner, royal gardener at Oreanda who assisted K. Koch during his travels in the Crimea.

Roemeria, called after Johann Jakob Roemer (1763–1819) of Zürich, editor of *Magazin für die Botanik.*

Rōmulea, called after Romulus, the legendary founder of Rome.

Rorippa, latinized form of old Saxon name *Rorippen.*

Rosa, the Latin name of several kinds of rose.

Rosea, substantive, former generic name, from *roseus, a, um,* rose-coloured, rose-like.

rostellātus, a, um, furnished with a little beak (*rostellum,* diminutive of *rostrum,* beak).

rubellus, a, um, reddish (diminutive of *ruber,* red).

rubens, pres.part. of *rubeo,* be red, blush.

ruber, bra, brum, red.

Rubia, name of madder (*Rubia tinctorum*) in Pliny.

rubicundus, a, um, ruddy (*rubeo*, be red).

rūbīginōsus, a, um, rusty, from the glandular lower surface of the leaves (better spelt *rō-*, from *rōbīgo, inis,* rust).

Rubus, the Latin word for brambles.

Rudbeckia, called after Olof Rudbeck, Proɪessor of Anatomy and Botany at Upsala, teacher and patron of Linnaeus.

rūderālis, e, growing in waste places or among rubbish (*rūdus, eris,* broken stones, old rubbish).

rūfescens, becoming red, reddish.

rūfus, a, um, red, reddish.

Rumex, name in Pliny, etc., usually translated sorrel.

rūpester, tris, tre, growing on rocks (*rupes,* rock, crag). Masculine -*tris* usual.

rūpicola, substantive, growing on rocks (*rūpes,* crag, and *colo,* inhabit).

Ruppia, called after H. B. Rupp (1688–1719), Central European botanical author.

rūrivagus, a, um, rambling about the country (*rūs, rūris,* country, and *vagus,* wandering).

Ruscus, *ruscum,* name of a prickly plant, perhaps *R. aculeatus.*

rusticānus, a, um, of the country (*rūs*).

Rūta-bāga, Swedish dialect name.

Rūta-mūrāria, name, in Brunfels, of a plant resembling RUTA spp. and growing on walls (*mūrus,* wall).

rutilus, a, um, red (inclining to golden yellow).

sabaudus, a, um, of Savoy (*Sabaudia*).

sabulicola, substantive, a dweller on sand (*sabulum,* sand, and *colo,* inhabit).

sabulōsus, a, um = *sabulicola,* original meaning, sandy.

saccharātus, a, um, sugared, the fruits of *Galium saccharatum* bear numerous wart-like projections and have the appearance of being sprinkled with sugar (σάκχαρον).

saccharīnus, a, um, adjective from σάκχαρον, sugar. *Acer saccharinum* (= *A. dasycarpum*) is not the sugar maple. The *i* is long if the adjective be considered Latin.

Sagīna, Latin word meaning food, fodder. *Spergula arvensis,* which was cultivated as a fodder crop, was formerly called *Sagina Spergula.*

Sagittāria, *sagittārius, a, um,* arrow-like, from the form of the leaves of *S. sagittifolia* (*sagitta,* arrow).

salebrōsus, a, um, rough.

71

salicārius, a, um = *salicīnus*.

salicīnus, a, um, willow-like (word made from *salix*).

Salicornia, from *sal*, salt, and *cornu*, horn, from the habitat and horn-like shoots.

salignus, a, um, willow-like, of willow (*salix*).

salīnus, a, um, growing in soil impregnated with salt (*sal*, salt).

salisburgensis, e, of Salzburg (*Salisburgia*), the type locality of *Euphrasia salisburgensis*.

Salix, the Latin name of various willows.

Salsola, name used by Cesalpino, from *sal*, salt, referring to the salt taste of *S. sativa* (*Halogeton sativus*).

saltuum, gen.pl. of *saltus*, woodland pasture.

Salvia, name of *S. officinalis*, perhaps from *salvus*, safe, sound, *salūs*, health, referring to healing properties.

Sambūcus, name of the elder-tree, perhaps so called from its numerous straight, erect, parallel, epicormic shoots, which resemble the strings of the musical instrument σαμβύκη (Rich, p. 571).

Samolus, name of a plant in Pliny.

sanguineus, a, um, blood-red (*sanguis, inis,* blood).

Sanguisorba, name first used by Fuchs, from *sanguis*, blood, and *sorbēre*, to swallow up, absorb. The root-stock is rich in tannin, and is a good styptic.

Sānicula, from *sāno*, heal, referring to its medicinal properties.

Santalāceae, called after the genus SANTALUM. Sandal-wood oil is distilled from the wood of *S. album*. (Medieval Latin, *sandalum*, of Arabic origin.)

Santolīna, *sanctum līnum*, holy flax, an old name of *S. virens*.

Sanvitalia, called after the Italian noble family Sanvitali.

Sāpōnāria, from *sāpo, ōnis*, soap. *S. officinalis* contains saponin and froths when rubbed up with water.

saracēnicus, a, um (sarracēnicus), pertaining to the *Saracēni* (Saracens), a people of Arabia Felix.

sardōus, a, um, Sardinian. *Herba sardōa* was the name of a poisonous plant, thought to be a RANUNCULUS.

sarmentāceus, a, um, having long, slender runners (*sarmentum*, twigs, brushwood, in botanical Latin a runner).

Sarothamnus, σάρον, broom, and θάμνος, shrub, referring to the habit of our species. See *scōpārius*.

Sarracenia, called after Dr D. Sarrasin of Quebec, who sent *S. purpurea* to Tournefort.

satīvus, a, um, sown, planted, cultivated (*sero*, sow), as opposed to *agrestis* and *silvestris*, which mean wild.

72

Satureia (Satureja), name, in Latin authors, of a fragrant potherb.

Saussurea, called after H. B. de Saussure (1779–96), the Swiss philosopher, author of *Voyages dans les Alpes*.

saxātilis, e, growing on or among rocks (*saxum*, rock).

Saxifraga, *saxifragus, a, um*, stone-breaking, adjective applied to herbs that grew among rocks, and also to those that had the reputation of dissolving stone in the bladder (*saxum*, rock, *frango*, break).

scaber, ra, rum, rough, scurfy.

scaberrimus, a, um, superlative of *scaber*, rough, scurfy, scabrous.

Scabiōsa, from *scabiōsus, a, um*, rough, scurfy, probably referring to the stem of *S. arvensis*, now called *Knautia arvensis*, and to its use for scabies.

scabriusculus, a, um, diminutive of the adjective *scaber*, rough, scurfy, meaning slightly rough.

scabrōsus, a, um, = *scaber*.

scandicus, a, um, of Schonen (*Scania*).

Scandix, σκάνδιξ, name of *S. pecten-veneris* in several authors.

Scariola, *Serriola*, several other forms occur in herbals. Diminutive of *seris*, chicory, etc.

scelerātus, a, um, wicked, vicious. *Ranunculus sceleratus* is exceedingly acrid, blisters and ulcerates the skin, and was formerly used by malingering beggars. 'Planta corrosiva exulcerat pedes mendicorum, ad nundinas se praeparantium' (Linnaeus, *Flora Suecica*).

Schedonorus, σχεδόν, near, and ὄρος, boundary. The awn of some species is attached near the apex of the lemma.

Scheuchzēria, called after the two brothers Scheuchzer, Swiss botanists.

Schkuhria, called after Christian Schkuhr (1741–1811), of Wittenberg.

Schoenoplectus, σχοινόπλεκτος, plaited of 'rushes', from σχοῖνος, and πλέκω, plait.

Schoenoprasum, σχοῖνος, kind of rush-like plant (see SCHOENUS), and πράσον, leek.

Schoenus, σχοῖνος, name of rush-like plants.

sciaphilus, a, um, shade-loving (σκιά, ᾶς, shade, and φίλος, friend).

Scilla, σκίλλα, the name of *Urginea Scilla* (= *U. maritima*). '*Scilla*' ('squill') of the *British Pharmacopoeia* is the bulb of this species divested of its dry membranous outer scales, cut into slices, and dried.

Scirpus, name of a rush-like plant, also written *sirpus*.

sciūroīdes, σκίουρος, squirrel, and *-oides*, like, referring to the long, tail-like panicle.

Sclēranthus, σκληρός, hard, and ἄνθος, flower, from the indurated perianth.

Sclērochloa, σκληρός, hard, and χλόη, grass.

Scleropoa, σκληρός, hard, and πόα, grass, from the hard texture of the species.

Scolopendrium, σκολοπένδριον, a name, in Dioscorides, of a plant resembling a millipede (σκολόπενδρα).

Scolymus, σκόλυμος, name, in Greek authors, of *S. hispanicus*.

scōpārius, a broom made of twigs (*scōpae*).

Scordium, σκόρδιον, name, in Dioscorides, of a plant smelling like garlic (σκόρδον).

Scorodonia, from σκόροδον or σκόρδον, garlic, plant resembling *Teucrium scordium*.

Scorodoprasum, Scordoprasum, σκορδόπρασον, name in Dioscorides of a plant resembling both the leek (πράσον), and the garlic (σκόρδον).

scorpioīdēs, resembling the tail of a scorpion (σκορπίος). See Arber, fig. 127, p. 253. The left-hand plant in this figure resembles a MYOSOTIS.

Scorzonēra, a name frequently used in herbals, perhaps from Italian *scorzone*, adder, because the plant was used for snake-bites.

Scrōphulāria, from scrōfulae, i.e. tubercular glands in the neck. The rhizome of *S. nodosa* resembles 'scrofulae'. See Arber 250–5.

scūtātus, a, um, adjective from *scūtum*, shield, originally meaning armed with a shield, now applied to plants with shield-like leaves.

Scutellāria, from *scutella*, salver, dish (Rich, p. 589), referring to the pouch on the calyx.

Secāle, the Latin name of a kind of grain, by some thought to be rye.

secālīnus, a, um, resembling rye (*secāle*).

secundus, a, um, arranged on one side only. Classical meaning, following.

Sedum, name of a plant, in Pliny, probably of SEMPERVIVUM sp.

segetālis, e, of cornfields (*seges, etis*, corn, crop).

segetum, of cornfields (gen.pl. of *seges*, corn, crop).

Selāginella, diminutive of *selāgo*.

Selāgo, name, in Pliny, of a plant resembling *Sabīna herba* (*Juniperus Sabīna*).

Selīnum, σέλῑνον, name of a plant, perhaps celery, used for making garlands. At least three derivations have been suggested, see Hooker, p. 181, and Hegi, v, 2, p. 1309.

sēmi-, half.

sēmidecandrus, a, um, hybrid word meaning with half (*sēmi-*) ten (δέκα) stamens (ἀνήρ, ἀνδρός, man). See *sēmi-*, *deca-*, and *-andrus*.

sempervirens, evergreen (*semper*, always, and *vireo*, be green).

Sempervīvum, name, in Pliny, of the house-leek, from *semper*, always, and *vivo*, live, referring to its retention of vitality. See *aīzoides*.

74

Senebiera, called after Jean Senebier (1742–1809), Swiss physiologist.

Senecio, name, in Pliny (see ERIGERON), probably of *Senecio vulgaris,* from *senex,* old man, because of the conspicuous white pappus.

senescens, growing old or hoary, referring to the white hairs on the tips of the phyllaries of *Hieracium senescens* (pres.part. of *senesco,* grow old; *senex,* old man).

senticōsus, a, um, thorny (*sentēs,* thorny shrubs of various kinds).

sēpincola, subst., a dweller in hedges, from *sēpes* (*saepes*), hedge, and *incola,* inhabitant.

sēpium, of hedges (gen.pl. of *sēpes* (*saepes*), hedge).

sept-, septem, seven.

septentriōnālis, e, northern.

Sequoia, from Sequoiah (1770–1834), inventor of Cherokee alphabet.

Serāpias, σεραπίας, name, in Pseudo-Dioscorides, of *Orchis longicruris.* The genus SERAPIAS, as now constituted, is Mediterranean.

sēricans = *sēriceus.*

sēriceus, a, um (better spelling *sēricus*), silky (*Sēres,* the Chinese, who were celebrated for their silken fabrics).

Seridia. See *-seris.*

Serīphidium, from σέρῐφον, name, in Dioscorides, synonymous with σαντονικόν, which was probably *Artemisia maritima.*

-seris, σέρις, ιδος, a kind of endive, suffix meaning potherb.

sērōtinus, a, um, late (in leaf or flower); note short *i* (*sēro,* adv., late).

Serpyllum, sibilated neuter form of ἕρπυλλος, name of *Thymus Sibthorpii* (? ἕρπω, creep).

Serrafalcus, called after Domenico Lo Faso Pietrasanta, Duca di Serrafalco, famous archaeologist.

Serratula, name, in Pliny, of a plant called also *Vettonica* (see *Betonica*). The word probably refers to serrate foliage.

Serriola, see *Scariola.*

Seseli, σέσελι, σέσελις, name of a plant in Greek authors, thought to be a species of TORDYLIUM.

Sesleria, called after Leonardo Seslero, Venetian doctor and naturalist, contemporary of Linnaeus.

sētāceus, a, um (**saetāceus**), bristly (*sēta, saeta,* bristle).

Sētāria, *sēta, saeta,* bristle, from the bristles on the peduncle of the spikelet.

sētifer, a, um, bristle-bearing (*sēta,* better spelling, *saeta,* and *fero,* bear).

sētiger, a, um = *sētifer* (*gero,* bear).

sētōsus, a, um = *sētāceus.*

Shallon, probably related to the Chinook jargon word *sallal* (Chinook, *kl-kwu-shá-la*), name of *Gaultheria Shallon.*

75

Sherardia, called after William Sherard (1659–1728), and his brother James.

Sibbaldia, called after Robert Sibbald (1643–1720), Professor of Medicine at Edinburgh, author of *Scotia illustrata.*

Sibthorpia, called after Humphrey Sibthorp, father of John Sibthorp, who was the author of *Flora Graeca.*

siculus, a, um, belonging to Sicily (*Sicilia*).

Siegesbeckia, called after Johann Georg Siegesbeck of Leningrad, botanical author.

Sieglingia, called after Professor Siegling, who botanized in the region of Erfurt early in the nineteenth century.

Silaüs, name of a plant in Pliny.

Silēne, origin obscure; see Leunis, p. 587, note 1.

silvāticus, a, um, growing in woods (*silva*, woodland).

silvester, tris, tre (masculine -*tris* more usual), in botanical Latin this word always means wild (*silva*, woodland).

Silybum, σίλυβον, name in Dioscorides of a kind of thistle, thought to be *Silybum Marianum.*

Simethis, origin obscure.

sīmia, an ape, from the form of the flower.

simplex, -icis, simple, unbranched.

simpliciusculus, a, um, diminutive of *simplex.*

Sināpis, *sināpis,* collateral form of *sināpe* and *sināpi* (σίναπι), mustard.

Sināpistrum. See SINAPIS and -*aster.*

sinuātus, a, um, with a wavy margin (p.part. of *sinuo*, bend; cf. *sinus,* a curve).

Sisōn, σίσων (σίνων), name, in Dioscorides, perhaps of *Sison Amomum.*

Sisymbrium, σισύμβριον (σίσυμβρον), name of various species of *Labiatae* and possibly *Cruciferae.*

Sisyrinchium, σισυριγχίον, name, in Theophrastus, of a plant with sweet tubers, said to be *Iris Sisyrinchium.*

sitchensis, e, from Sitka, in Alaska.

Sium, σίον, name of at least two water-plants, thought to be *Sium angustifolium* (*Berula erecta*) and *Veronica Anagallis.*

Smyrnium, σμύρνιον, name of *Smyrnium perfoliatum,* from σμύρνα, myrrh, because of the aroma.

Sōlānum, name of a plant in Pliny and Celsus.

Soldanella, perhaps from *soldo,* Italian coin, referring to the orbicular leaves of SOLDANELLA (*Primulaceae*) and *Calycostegia Soldanella.* According to Hegi, however, the name was also applied to *Lycopodium clavatum.*

Solidāgo, name first used for BELLIS by Brunfels, from *solido,* strengthen, fasten together, from being used as application to wounds. The feminine suffixes *-āgo* and *-ūgo* are common in plant-names.

solstitiālis, flowering at the time of the summer solstice (*solstitium*).

somnifer, era, erum, sleep-bringing, soporific. *Papaver somniferum* yields opium (*somnus,* sleep, and *fero,* bear).

Sonchus, σόγχος, σόγκος, the name of species of SONCHUS and perhaps of other thistle-like plants.

Sophia, σοφία, wisdom, *Sophia chirurgorum* (wisdom of the surgeons), name in herbals of *Descurainia* (*Sisymbrium*) *Sophia,* from its use in healing.

Sorbus, the Latin name of *Sorbus domestica.*

spādiceus, a, um, Latin adj. formed from σπάδιξ, ίκος, date-coloured.

Sparganium, σπαργάνιον, name of a plant, probably *Sparganium simplex,* in Dioscorides, diminutive of σπάργανον, band for swathing infants, in plural, swaddling clothes. The plant is perhaps so called from the form of the leaves.

Spartīna, *spartum,* σπάρτον, name of various plants used for making ropes, etc. Best regarded as a Latin word with a long *i.*

spathulātus, a, um, of leaves, etc., spathulate, oblong with the base gradually attenuated and the apex obtuse.

speciōsus, a, um, showy, handsome.

spectābilis, e, admirable, conspicuous (*specto,* look at).

Speculāria, from *speculum,* mirror. *S. Speculum* and *S. hybrida* were formerly known as *Speculum Veneris* and *Speculum Veneris minus.*

Spergula, word used by de l'Obel, probably formed from a German name.

Spergulāria, word formed from SPERGULA.

Spermophyta, neut.pl.subst. from σπέρμα, seed, and φῦτόν, τό, plant. Note the short *y.* The stem of σπέρμα is σπέρματ-, but the shortened stem σπερμο- was used in classical Greek (e.g. σπερμολόγος, picking up seeds).

-spermus, -a, -um, -seeded (σπέρμα, seed).

sphaerocephalus, a, um, round-headed (σφαῖρα, ball, and κεφαλή, head).

sphaerospermus, a, um, σφαῖρα, ball, and σπέρμα, seed. The achenes of *Ranunculus sphaerospermus* are almost spherical.

sphēgōdēs, σφήξ, σφηκός, wasp, and -ώδης, having the form of (*g* for *k* possibly euphonic).

Sphondylium, σφονδύλιον (*spondylium* in Pliny), name, perhaps, of *Heracleum Sphondylium,* σφόνδυλος = *verticillum.* See *verticillatus.*

Spīca-ventī, *spica,* ear, spike, and *venti,* genitive of *ventus,* wind. Name probably invented by Linnaeus, of doubtful meaning.

Spĭlanthēs, σπῖλος, spot, stain, and ἄνθος, flower, said to refer to the markings on the disk of some species.

spĭnōsus, a, um, prickly, thorny (spīna, thorn).

Spĭraea, σπειραία, name, in Theophrastus, of a shrub, thought to be *Ligustrum vulgare*, perhaps applied to our genus on account of the spirally twisted fruits (σπεῖρα, anything twisted or wound).

Spĭranthēs, σπεῖρα, anything twisted, and ἄνθος, flower, from the twisted inflorescence.

Spĭrodēla, σπεῖρα, coil, and δῆλος, obvious, manifest, from the distinct spiral vessels.

sponhemicus, a, um, of Sponheim (*Sponhemium*), Rhineland castle.

-sporus, a, um, -spored, -seeded (σπόρος, seed).

spurius, a, um, false, spurious.

squālidus, a, um, rough, unkempt, neglected, squalid.

Squāmāria, from *squāma*, scale, referring to the scale-clad root-stock (*rādix squāmāta*).

squāmātus, squāmōsus, a, um, scaly (*squāma*, scale).

squarrōsus, a, um, scurfy, with spreading processes of any kind (perhaps a corruption of *squāmōsus*).

Stachygynandrum, -stachyon, q.v., γυνή, female, and ἀνήρ, ἀνδρός, male, referring to micro- and macrosporangia in the same strobilus. Name used when SELAGINELLA was included in LYCOPODIUM.

-stachyon, -stachyus, a, um, -stachys, -spikeleted, -panicled (στάχυς, ear of corn, used in botany for spikelet and other narrow inflorescences or arrangement of sporangia).

Stachys, στάχυς, name of *Stachys germanica* in Dioscorides.

stagnīnus, a, um (botanical Latin), growing in swampy ground (*stāgnum*, pond, swamp).

Staphylēa, name, formed by Linnaeus, from σταφυλή, bunch of grapes, referring to the form of the inflorescence.

Staticē, στατική, name in Dioscorides of a herb, so called from being used as an astringent (στατικός, causing to stand, astringent).

Steironēma, στεῖρος, barren, and νῆμα, thread, referring to the staminodes.

Stellāria, medieval name, applied to plants with stellate arrangement of leaves or star-like flowers (*stella*, star).

stellāris, e, star-like (*stella*, star).

stellātus, a, um, with spreading rays, like those of a star.

stelliger, a, um, star-bearing (*stella*, star, and *gero*, bear).

-stēmōn, -stamened, στήμων = the Latin *stāmen*. The original meaning was the warp in an ancient upright loom (Rich, p. 620).

steno-, narrow- (στενός).

78

stenophyllus, a, um, having narrow leaves. στενός, narrow, and φύλλον leaf.

sterilis, e, barren, sterile. *Anisantha sterilis* is so called because of its rapidly falling flowers, *Potentilla sterilis* (*Fragaria sterilis* of Linnaeus) because of its dry *fruit*, unlike a strawberry.

stictophyllus, a, um, having spotted leaves (στικτός, spotted, and φύλλον, leaf).

Stramonium, origin obscure, said to be a corruption of στρύχνος μανικός, name, in Theophrastus, of *Datura Stramonium*.

Stratiōtēs, στρατιώτης, soldier, from the sword-like foliage. στρατιώτης ποτάμιος was the name in Dioscorides of an Egyptian water-plant (PISTIA).

striātus, a, um, striate, with faint ridges or grooves forming longitudinal, parallel lines (*stria*, flute of a column).

strictus, a, um, drawn together, erect, straight (p.part. of *stringo*).

strigōsus, a, um, in botanical Latin, beset with sharp-pointed stiff hairs or bristles.

strigulōsus, a, um, diminutive form of *strigōsus*, meaning slightly scaly, etc.

Strobus, name of a coniferous tree in Pliny (στρόβιλος, fir cone).

strūmārius, a, um, used in the treatment of *struma*, i.e. tubercular glands in the neck, which the involucres of *Xanthium strumarium* resemble.

stȳlus, in compound words, στῦλος, style, originally meaning pillar.

Suaeda, from the Arabic name of *S. baccata*.

suāveolens, sweet-scented (*suāvis*, sweet, agreeable, and *oleo*, to emit a smell).

sub-, under, in minor degree, almost; but when affixed to the terms of number, as in *subuniflorus*, it signifies most commonly.

sublustris, e, having a faint light, glimmering (*sub-*, q.v., and *lustro*, light up).

submersus = *dēmersus*, q.v. (better spelt *summ-*).

Sūbulāria, from *sūbula* (Rich, p. 629), awl, from the form of the leaves.

succīsus, a, um (p.part. of *succīdo*), cut off from below, referring to the curiously truncated rhizome of *Succisa pratensis*, which was supposed to have been bitten off by the Devil (*Rādix morsus diaboli*).

Succowia, called after Georg Adolph Suckow (1751–1813), Professor at Heidelberg, author of several botanical works.

suecicus, a, um, Swedish.

suffōcātus, a, um, strangled, suffocated. The sessile heads of *Trifolium suffocatum* are often half-buried in the soil.

sulcātus, a, um, grooved (*sulcus*, groove).

supīnus, a, um, lying on the back, stretched out, extended.

surculosus, a, um, woody, (in botanical Latin) suckering.

Symphoricarpus, from συμφέρω, bear together, and καρπός, fruit, from the clustered berries.

Symphytum, σύμφυτον, name of 2 plants in Dioscorides, so called from their use in healing wounds (συμφύω, make to grow together).

syn- (**sym-** before *b*, *m* and *p*), united (σύν, with, together).

sylv-. See *silv-*.

Sȳringa, σύριγξ, ιγγος, pipe or tube. So called by Dodoens because flutes can be made of the stems.

Tabernaemontāni, of Tabernaemontanus (latinized form of Bergzabern), author of a famous herbal (Arber, p. 76).

Tacamahacca, originally an Aztec word.

Tamarix, a late Latin word, for *tamariscus*. Origin obscure.

Tamus, perhaps a wrong spelling of *tamnus*, the name, in Pliny, of a wild vine which bore fruits called *taminia uva*.

Tanacētum, from the medieval Latin name *tanazita*, ultimately from ἀθανασία, immortality (ἀ privative, and θάνατος, death). 'Cadaver si ea fricetur, non a vermibus infestatur' (Linnaeus, *Flora Suecica*).

Taraxacum, medieval Latin, ultimately from Persian *tal̲k̲h̲ chak̲ō̲k*, bitter potherb. See art. in *Oxford English Dictionary*.

taurīnus, a, um, of Turin (*Augusta Taurinorum*). The word also means of or belonging to bulls (*taurus*, bull).

Taxōdium, *taxus* and *-ōdēs*, q.v., referring to the supposed resemblance to the yew.

Taxus, the Latin name of the yew.

tectōrum, of, or growing on, roofs (gen.pl. of *tectum*, roof).

Teesdālia, called after Robert Teesdale, a Yorkshire botanist.

Tēlephium, τηλέφιον was the name of a plant thought to be *Andrachne telephioides*: τηλέφιλον (τῆλε, far off, and φίλος, lover) was the leaf of some plant used as a charm by lovers to try whether their love was returned (see Liddell and Scott).

Telmatēia, τελματιαῖος, of a marsh (τέλμα, ατος, marsh).

Telmatophacē, τέλμα, ατος, pool, and φακῇ, lentil. φακὸς ὁ ἐπὶ τῶν τελμάτων (lentil on the pools) was the name of a plant in Dioscorides. The bulging aerenchyma of *Lemna gibba* gives the frond a faint resemblance to a lentil.

tēmulentus, a, um, drunken. Poisoning by *Chaerophyllum temulentum* and *Lolium temulentum* may simulate alcoholic intoxication. The word is often mis-spelt *temulus*.

tenellus, a, um, somewhat tender or delicate (dim. of *tener*, tender).

tenuis, e, fine, slender.

tephrosanthos, badly formed from τεφρός, ash-coloured, and ἄνθος, flower. Should be *tephranthus*.

teres, etis, terete, cylindrical (lit. rounded off, from *tero*, rub).

teretiusculus, a, um, diminutive of *teres, -etis,* smooth, rounded; in botany, cylindrical. *Carex teretiuscula* has weakly triangular stems.

terrester, tris, tre (masculine usually **-tris**), belonging to the earth, terrestrial (*terra*, earth).

tetra-, τέτρα-, four-.

Tetracmē, *tetra-*, four-, and ἀκμή, point, from the four cusps on the fruits.

Tetragōnolobus, from τετράγωνος, quadrangular, and λοβός, pod.

Tetrahit, old name of *Pimpinella Tragium,* afterwards transferred to various *Labiatae* (perhaps connected with *teter,* foetid, referring to the goat-scented rootstock).

Tetralix, τετράλιξ, name of a thistle-like plant in Theophrastus.

tetraplus, a, um, τετραπλόος, fourfold, having leaves arranged in 4 ranks.

Teucrium, τεύκριον, name of a plant in Dioscorides, called after the hero Teucer (Τεῦκρος).

Teutliopsis, from τευτλίον, diminutive of τεῦτλον, BETA, and ὄψις, resemblance.

Thaliānus, a, um, of Johannes Thal, sixteenth-century German physician.

Thalictrum, θάλικτρον in Dioscorides was probably a plant of the *T. minus* aggregate.

Thapsus, θάψος, name of *Cotinus Coggygria,* used for dyeing yellow (from island of Thapsos).

Thēlypteris, θηλύπτερις, name of a fern in Theophrastus, from θῆλυς, female, and πτερίς, fern; cf. *Filix-femina,* 'lady fern'.

Thēsīum, θησεῖον, name of a bulbous plant in Pliny.

Thlaspi, θλάσπις, name, in Hippocrates (θλάσπι in Dioscorides), of a plant whose seeds, when crushed (θλάω), were used like mustard.

Thrauosphaereae, should be *Thraustosphaereae,* from θραυστός, brittle, and σφαῖρα, ball, sphere, here meaning pollinium.

Thrincia, from θριγκός, coping, referring to the form of the pappus of the outer florets, which consists of toothed scales.

Thuja, θυία (θύον), name, in Theophrastus, of a tree with fragrant wood, probably *Tetraclinis articulata.*

Thymelaeaceae, called after the genus THYMELAEA; θυμελαία was the name, in Dioscorides, of *Daphne Gnidium.*

Thymus, θύμος, name of a plant, probably *Corydothymus capitatus,* in Theophrastus, etc., perhaps so called because it was used in offerings (θύω, sacrifice).

thyrsiflōrus, a, um, with flowers arranged in a contracted panicle or thyrse (θύρσος, bacchic staff (Rich, p. 662)).

Tilia, the Latin name of the TILIA spp.

Tillaea, called after Michelangelo Tilli (1655–1740), Professor of Botany at Pisa.

tinctōrius, a, um, used in dyeing (*tingo*, dye).

tinctōrum, of dyers (gen.pl. of *tinctor*, dyer).

Tinea, called after Vincenzo Tineo (1791–1856), Professor of Botany at Palermo. The Latin word *tinea* means moth.

Tiniāria, origin obscure, perhaps from *tinea*, moth.

Tinus, the Latin name of *Viburnum Tinus*.

Tithymālus, τιθύμαλλος, name, in classical authors, of various plants with milky juice, especially EUPHORBIA spp.

Tofieldia, called after Tofield, a Yorkshire botanist.

Tolpis, meaningless name coined by Adanson.

tōmentōsus, a, um, covered with hairs which are matted together to form a *tōmentum*, or blanket (lit. stuffing).

Tordylium, τορδύλιον, name, in Dioscorides, of a plant, thought to be *Tordylium apulum*.

Torilis, word made by Adanson, whose names are often meaningless.

Tormentilla, medieval Latin word meaning torment. The powdered rhizome of *Potentilla erecta* is rich in tannic acid and was used for various disorders.

torminālis, e, good for colic (*tormina*).

Trachēlium, word first used by Dodoens, from τράχηλος, neck, from the use of *Campanula Trachelium* in affections of the throat.

trāchy-, τραχύς, rough.

trāchyodon, with rough teeth, τραχύς, rough, and ὀδούς, ὀντος, tooth.

Trachystēmōn, τραχύς, rough, and στήμων (in botany), stamen, from the hairy filaments of *T. orientalis*.

Tragopōgōn, plant-name, in Theophrastus, τραγοπώγων, from τράγος, goat, and πώγων, beard, perhaps referring to the conspicuous pappus.

Tragus, τράγος, he-goat, also a plant-name in Dioscorides.

transiens, passing over, becoming transformed, hence used for species intermediate between other species (pres.part. of *transeo*, cross over).

transwalliānus, a, um, of the site of an English colony in Pembroke-shire, formerly called *Anglia Transwalliana* (Little England beyond Wales).

traunsteinerioīdēs, resembling *Orchis Traunsteineri*, called after Joseph Traunsteiner (1798–1850), pharmacist at Kitzbüchel in the Tirol.

tremulus, a, um, shivering, trembling (*tremo*, quake, quiver).

trī, three- (from *trēs*, three).

trĭ-, three- (τρĭ-, stem of τρεῖς, three).

tricho-, hairy, hair-like (θρίξ, τρῐχός, hair).

trichocaulon, having a hairy stem, Greek neut.adj. formed from θρίξ, τριχός, hair, and καυλός, stem.

Trichōdium, θρίξ, τριχός, hair, and -ώδης, of the nature of, referring to the very narrow radical leaves.

Trichomanes, τριχομανές, name of a plant in Theophrastus, thought to be *Asplenium Trichomanes.* Origin obscure.

Trichonēma, θρίξ, τριχός, hair, and νῆμα, thread, referring to the filiform style.

Trichophorum, θρίξ, τριχός, hair, and φέρειν, bear, from the long perianth bristles.

trichophyllus, a, um, having leaves divided into hair-like segments (θρίξ, τριχός, hair, and φύλλον, leaf).

tricornis, e, with three horns (*trī-,* three, and *cornū,* horn).

tridactylītēs, three-fingered, from τρεῖς, three, δάκτυλος, finger, and the suffix *-ίτης.*

Trientālis, *Herba trientālis* in Cordus, meaning herb the third of a foot high (*trientālis, e,* containing the third of a foot).

trifidus, a, um, three-cleft. Note the short first *i.* (The word is from *terfindo,* not *trifindo.*)

Trifolium, *ter,* thrice, and *folium,* leaf. Name, in Pliny, of plants with 3-foliate leaves.

Triglōchin, τριγλώχις, three-barbed, from the form of the dehiscing fruit of *T. palustris* (γλωχίν, barb of an arrow).

Trigōnella, feminine Latin diminutive of τρίγωνος, triangular. The carina of *T. Faenum-graecum* is so inconspicuous that the corolla appears to consist of three almost equal petals.

Trinia, called after Karl Bernhard Trinius (1778–1844), Russian botanist.

Triodia, badly ormed word, from τρι-, three, and ὀδούς, tooth, from the 3-toothed lemmata.

Trionychon, τρι-, three, and ὄνυξ, υχος, nail, claw, referring to the one bract and two prophylls.

Tripolium, τριπόλιον, name of a plant in Theophrastus, perhaps from τρι-, three, and πόλιον (*Teucrium Polium*), meaning three times as powerful as that plant.

triquetrus, a, um, having three acute angles.

Trisētum, *tri-,* three, and *saeta,* bristle, referring to the 3-awned lemmata.

Trīticum, wheat (*tero, trītum,* rub, thresh).

triviālis, e, belonging to the cross-roads, common, found everywhere (*trivium,* place where three roads meet; see Rich, p. 693).

Trollius, name first used by Gesner, from the German *Trollblume.*

Tsuga, said to be the Japanese name of *T. Sieboldii.*

Tüberāria, from *tūber*, swelling, tuber. The type species, *T. vulgaris*, has a thick woody root-stock.

Tulipa, ultimately from a Persian word for turban, to which the flowers were likened.

tumidicarpus, a, um, *tumidus*, swollen, protuberant, and καρπός, fruit. See *lepidocarpus*.

Tunica, name of doubtful origin perhaps referring to a plant used medicinally in Tunis.

Turgenia, called after Alexander Turgeneff, Director of the Chancellery of Prince Gollintzin at Moscow.

Turrīta, Turrītis, a plant-name used in herbals, said to be derived from *turris*, tower, either from the form or, more probably, from the habitat of the plant.

Tussilāgo, name of a plant in Pliny, from *tussis*, cough, referring to its medicinal uses, and feminine suffix *-āgo*.

Tÿpha, τύφη, name of several plants in Greek authors.

-typus, suffix denoting typical (τύπος, archetype, pattern, model).

Üdōra, name badly formed from ὑδώρ, water, or Latin *ūdor*, moisture, referring to the habitat.

Ulex, name, in Pliny, of a shrub resembling Rosemary.

ūlīginōsus, a, um, growing in marshy places (*ūlīgo, inis*, moisture, marshy quality).

Ulmāria, word invented by Gesner, and meaning elm-like, referring to the leaflets of *Filipendula Ulmaria* Maxim. (*Spiraea Ulmaria* L.).

Ulmus, the Latin name of ULMUS spp. (cognate with English 'elm').

Umbelliferae, fem.pl. (sc. *plantae*) of *umbellifer, a, um*, umbel-bearing, formed from *umbella*, diminutive of *umbra*, shadow, shade, sunshade, parasol, umbel, and *fero*, bear.

Umbilīcus, the navel, referring to the depressed centre of the leaves. *U. pendulinus* (*U. rupestris*) was called *Umbilicus Veneris* in the fifteenth-century herbal called *Herbarium Apuleii Platonici*.

umbrōsus, a, um, growing in shady places (original meaning, shady, from *umbra*, shade, shadow).

uncinātus, a, um, furnished with a hook (*uncus*).

uncinellus, a, um, furnished with a small hook (*uncus*, hook).

undulātus, a, um, wavy (*unda*, wave).

Ünedo, the fruit of ARBUTUS, and also the tree itself.

unguiculātus, a, um, furnished with a claw (*unguis*, claw; the diminutive *unguiculus* originally meant finger-nail).

ūniseriātus, a, um, arranged in one row, from *ūnus*, one, and *seriātus*, adjective formed from *series*, series, row.

urbānus, a, um, belonging to the city (*urbs*).

urbicus, a, um, belonging to the city (*urbs*).

ūrens, stinging, acrid (*ūro*, burn, sting).

Ūrostachya, οὐρά, tail, and *stachyon*, q.v. The sporangia are borne, not in a 'cone', but in a 'tail' in the axils of the upper leaves.

ursīnus, a, um, connected with bears. The combination *allium ursinum* occurs in Pliny (*ursus*, bear).

Urtīca, sting-nettle. Note the long *i* (*uro*, burn, sting).

ūsitātissimus, a, um, superlative of *ūsitātus*, customary, ordinary, familiar.

ustulātus, a, um, p.part. of *ustulo*, scorch, singe, from the appearance of the inflorescence.

Utriculāria, from *utriculus*, diminutive of *uterus*, belly, womb, from the insect-catching bladders.

Ūva-crispa, *ūva*, grape, and *crispus, a, um,* curly; translation, probably based on false etymology, of the German dialect word *Kräuselbeere*, which is of doubtful origin.

Ūva-ursī, *ūva*, grape, and *ursī*, gen. of *ursus*, bear.

Vaccāria, name first used by d'Aléchamps, probably from *vacca*, cow.

Vaccīnium, probably a corruption of *Hyacinthus*, ὑάκινθος; not connected with *vacca*, cow.

vagans, wandering (pres.part. of *vagor*).

vagensis, e, of the River Wye (*Vaga*).

vāgīnātus, a, um, sheathed, having conspicuous sheaths (*vāgīna*, sheath).

vagus, a, um, strolling, roving, uncertain, doubtful.

Valerandi, of Dourez Valerand, sixteenth-century botanist, who thought that *Samolus Valerandi* was the *samolus* of Pliny.

Valēriāna, medieval name, perhaps from *valeo*, be well.

Valerianella, diminutive of VALERIANA.

vallesiānus, a, um, of Valais (*Vallesia*), Swiss Canton.

Vallisnēria, called after Vallisnieri de Vallisnera (1661–1730), Professor at Padua.

variātus, a, um, manifold, various.

vectensis, e, of the Isle of Wight (*Vectis*).

vegetus, a, um, vigorous.

Vella, origin obscure.

velūtīnus, a, um, velvety.

Venīlia, name of several sea-nymphs.

ventricōsus, a, um, distended, inflated (*venter, tris,* belly).

vēnulōsus, a, um, with a network of numerous fine veins (*vēnula*, diminutive of *vēna*, vein).

85

Verbascum, name of a plant in Pliny.

Verbēna, a Latin word meaning foliage, twigs, etc., used for religious purposes.

Verbēnāca, name of a plant in Pliny.

vēris, of the spring (gen.sing. of *vēr,* spring).

Verlotōrum, gen. pl., called after two brothers Verlot, who first distinguished *Artemisia Verlotorum* from *A. vulgaris.*

vernālis, e, of (flowering in) spring (*ver,* spring).

vernus, a, um = *vernālis.*

Veronica, perhaps called after St Veronica. The name, which first occurs in Fuchs, may be a corruption of *Vettonica.* See *Betonica.*

verrūcōsus, a, um, full of warts, warty (*verrūca,* wart).

versicolor, changing colour (*verto,* change, and *color,* colour).

verticillātus, a, um, whorled. The *verticillus* was the whirl, whorl, or disk of a spindle. See Rich, p. 721.

vērus, a, um, true, genuine.

vescus, a, um, small, feeble; but in botanical Latin probably always means edible.

vēsīcārius, a, um, bladder-like, e.g. the perigynia of *Carex vesicaria* (*vēsīca,* bladder).

vespertīnus, a, um, of (flowering during) the evening.

vestītus, a, um, clothed, clad, covered with appressed hairs as with a garment (p.part. of *vestio,* clothe).

Vetrix, latinized form of the Italian word *vetrice,* osier.

vexans, annoying. The affinities of *Sorbus vexans* are very puzzling.

Viburnum, name of a plant, thought to be *V. Lantana.*

Vicia, name in Latin authors of a leguminous plant.

Vignea, called after Gislain François de la Vigne, early nineteenth-century botanist, Professor of Botany at Charkow, who made a French edition of Schkuhr's monograph of CAREX.

vīminālis, e, bearing twigs for plaiting (*vīmen, inis,* pliant twig, *vieo,* plait, weave).

vīnāceus, a, um, belonging to wine or to the grape (*vīnum,* wine).

Vinca, *vinca pervinca,* name of a plant in Pliny, whence our English name periwinkle, perhaps from *vincio,* wind, bind.

vīneālis, e, of or belonging to vines, from *vīnea,* vineyard. Compare *Ampeloprasum.*

Viola, Latin name (corresponding to ἴον) of *Viola odorata, Cheiranthus Cheiri,* and other plants with sweet-scented flowers.

virens, pres.part. of *vireo,* be green.

virgātus, a, um, with long, straight, slender shoots (*virga,* slender green shoot, wand (Rich, pp. 727–8), twig).

86

Virgaurea, name used in herbals, from *virga*, rod, and *aureus, a, um,* golden.

vīrōsus, a, um, poisonous, fetid, slimy.

Viscāria, from *viscum*, bird-lime, referring to the viscid stems of *V. vulgaris*.

viscōsus, a, um, sticky, viscid, lit. full of bird-lime (*viscum*).

Viscum, the Latin name for mistletoe and bird-lime.

Vītalba, name of our species of CLEMATIS, contracted from *vītis alba,* white vine.

vitellīnus, a, um, coloured yellow like the yolk of an egg (*vitellus,* yolk).

Vītis-Idaea, vine of Mount Ida, translation of ἄμπελος παρὰ Ἴδης, a plant-name in Theophrastus.

vītisalix, perhaps from *vītilis, e,* plaited, interwoven (cf. *vītilia,* wicker-work), and SALIX.

Vogelia, called after Christian Benedict Vogel (1745–1825), Professor at Altdorf, botanical author.

vulgāris, e, common.

vulgātus, a, um = vulgāris.

vulnerārius, a, um, used for wounds (*vulnus, eris,* wound).

Vulpia, called after Johann Samuel Vulpius (1760–1846), pharmacist in Pforzheim, who investigated the Flora of Baden.

vulpīnus, a, um, of, or belonging to the fox (*vulpes*). *Carex vulpina* has fox-coloured inflorescence.

Vulvāria, word first used by Durante, referring to the odour of the plant.

Wahlenbergia, called after G. Wahlenberg (1780–1851), Professor of Botany at Upsala.

Wolffia, called after J. F. Wolff (1778–1806), doctor in Schweinfurt, who wrote about LEMNA.

Woodsia, called after Joseph Woods (1776–1864), English botanist.

Xanthium, ξάνθιον, name, in Dioscorides, of *Xanthium strumarium,* which was used for dyeing the hair yellow (ξανθός, yellow).

xanth(o)-, yellow (ξανθός).

xanthochlōrus, a, um, yellowish green (ξανθός, yellow, and χλωρός, light green).

Xanthozōon, ξανθός, yellow, and ζῷον, living being. The word is perhaps formed on analogy with AÏZOON, and referring to the yellow flowers.

Xylosteum, ξύλον, wood, and ὀστέον, bone, referring to the hard wood.

Zannichellia, called after G. G. Zannichelli (1662–1729), chemist in Venice. His botanical works were published posthumously.

Zeobromus, ζέα (ζειά), a kind of grain, and BROMUS.

Zerna, ζέρνα = κύπειρος (CYPERUS sp.), referring to the form of the spikelets.

zetlandicus, a, um, of Shetland.

Zizii, of J. B. Ziz (1779–1829), botanist, teacher in Mainz.

Zöstëra, ζωστήρ, name of a marine plant in Theophrastus (probably *Poseidonia oceanica*). The word ζωστήρ means a belt or girdle.

APPENDIX

acūtus, a, um, sharp-pointed.

aequālis, e, equal, like, resembling. *Polygonum aequale* is homoio-phyllous, and *Alopecurus aequalis* resembles *A. geniculatus*.

Agapanthus, ἀγάπη, love, and ἄνθος, flower.

Agērătum, ἀγήρᾱτος, ον, ageless, not growing old, a plant that does not readily wither (ἀ, privative, and γῆρας, old age). The ἀγήρατον of Dioscorides was the name of several *Labiatae*.

Ailanthus, aylanto, the name in Amboyna of *A. moluccana*.

albicans (being) white.

aleppicus, a, um, of Aleppo.

Aleuritia, ἄλευρον, flour. The leaves are mealy beneath.

alkekengi, ἀλικάκαβον, name in Dioscorides of *Physalis alkekengi*.

alliāceus, a, m, better spelt with one *l*, smelling of garlic (*alium*).

Althaea, name in Theophrastus perhaps of *A. officinalis*, but see Hegi, v, i, 463, footnote (ἀλθαίνω, heal).

ambiguus, a, um, doubtful, of puzzling affinities.

amoenus, a, um, lovely, delightful, pleasant.

anceps (substantive), something with 2 heads; in botany, flattened, 2-edged.

Apium, see Hegi, v, ii, p. 1139, footnote.

aquilīnus, a, um, add: 'radix oblique dissecta refert aliquatenus aquilam imperialem' (Linnaeus, *Flora Suecica*).

Arctium, probably from ἄρκτος, bear, referring to woolly covering.

arēnastrum, probably badly formed word from ARENARIA and *-aster*, *-astrum*, q.v. *Polygonum arenastrum* resembles a depauperate ARENARIA.

Asarīna, so called from the leaves resembling those of ASARUM.

Asphodelus, ἀσφόδελος, the name of *Asphodelos ramosus*.

assimilis, adsimilis, e, resembling (some other species).

Athanāsia, see TANACĒTUM.

atlanticus, a, um, of Mount Atlas or its vicinity.

auriculae-ursifolius, see AURICULA.

benedictus, a, um, blessed (church Latin).

Brathys, βράθυ, name, in Dioscorides, of JUNIPERUS spp. Andean species of this section have juniper-like foliage.

brevipilus, a, um, with short hair (*brevis*, short, and *pilus*, hair).

89

brūmālis, e, of winter (*brūma*, winter solstice, from *brevissima* (*dies*), shortest (day)).

Buddleja, named after Adam Buddle, a vicar of Farnbridge.

bulbōsus, a, um, having a bulb or structure similar in appearance.

calēdonicus, a, um, of *Caledonia*, ancient name of North Britain.

Calodendron, καλός, beautiful, and δένδρον, tree. Name of a genus of *Rutaceae* now spelt -*um*.

cambrensis, e = *cambricus*.

Camelina, possibly derived from χαμαί, on the ground, dwarf, and λίνον, flax.

capitātus, a, um, with flowers in capitula; pin-headed, as stigmas, etc.

cāricus, a, um, of Caria, province in Asia Minor.

Caryophyllus, add: The early Arabic writers called cloves Karanfal and similar names derived from the language of the Malabar coast, Ceylon, etc.

cassubicus, a, um, of Cassubia, part of Pomerania.

Catabrōsa, add: or to the fact that cattle eat the plant with avidity.

catholicus, a, um, *Diplotaxis catholica* was known to Linnaeus only from Spain and Portugal, both Catholic countries.

Cerasus, probably from the Caucasian names kirahs, kiljas, of *Prunus avium*. The town in Pontus was called after the tree, not vice versa, see Hegi, IV, ii, 1074, footnote.

cespitōsus = *caespitōsus*.

chalapensis, e, of Aleppo (Arabic حلب).

Chamaemēlum, *see* CHAMOMILLA.

chylla, from the vernacular name of *Pinus longifolia* and other species of PINUS.

Cinnamomeae, called after *Rosa cinnamomea*.

colōrātus, a, um, coloured.

commūtātus, a, um, *Bromus commutatus* was formerly confused with *B. racemosus*.

cōnicus, a, um, conic, cone-shaped. The calyx of *Silene conica* scarcely justifies this epithet.

compactus, a, um, closely packed (p.part. of *compingo*).

conjunctus, a, um, joined together (p.part. of *conjungo*).

Cortadēria, American Spanish *cortadera*, the name of *C. Selloana*, first meaning a cutting instrument. The leaf margins and keels are spinous -serrulate.

Cracca, an Italian name of *Vicia cracca*. The word occurs as a plant name in Pliny.

crāvoniensis, e, of Cravon, district of Yorkshire.

Crīnĭtāria, from *crinitus*, long-haired, p.part. of *crinio*, referring to the inflorescence. See *Chrÿsocoma*.

Cyrtōmium, from κύρτωμα, bulge, swelling, referring to the form of the leaflets.

Dactylorchis, δάκτυλος, finger, referring to the palmate tubers.

dānicus, a, um, Danish.

diandrus, a, um, having 2 stamens (δίς, twice, doubly, and ἀνήρ, ἀνδρός, man, i.e. stamen).

diaphanoīdēs, a species allied to *Hieracium diaphanum*; both have diaphanous foliage.

Dichondra, δίς, twice, double, and χόνδρος, granule, lump, referring to the 2-lobed ovary.

dīlātātus, a, um, spread out, broad (p.part. of *dilato* from *lātus*, broad).

Dorycnium, δορύκνιον, name in Dioscorides of *Convolvulus oleaefolius*.

dulcis, e, sweet, also used for *mītis*, q.v.

Ecballium, from ἐκβάλλω, throw out. The fruit, when ripe, becomes detached and squirts the seed to a considerable distance. The plant thus bears a perfect 'signature' of a hydrogogue cathartic, which it is.

echīnosporus, a, um, ἐχῖνος, hedgehog, and σπόρος, seed. The macrospores are covered with sharp spines.

ēdentulus, a, um, without teeth. (*e* (*ex*), not having, and *dens, tis,* tooth).

Elatērium, ἐλατήριον, name in Greek authors of *Ecballium Elaterion*, from ἐλατήριος, driving away, ἐλατήρια φάρμακα, purgative medicines. See ECBALLIUM.

eleo, see *heleo*.

ērectus, a, um, upright (p.part. of *ērigo*, set up, erect).

Ērūca, the Latin name of *Eruca sativa*, still so called in Italian. Probably connected with *ēructo*, belch.

Ervum, name in Latin authors of *Vicia ervilia* = ὄροβος in Theophrastus and other Greek writers. See OROBUS.

exīlis, e, small, thin, meagre.

Fīcus, the Latin name of the fig tree and its fruit.

flaccus, a, um, flabby, pendulous.

-flōrus, a, um, -flowered (*flōs, flōris,* flower).

fluviātilis, e, in botany applied to plants growing in running water (*fluvius*, river, stream).

foliōsus, a, um, leafy (*folium*, leaf).

frondōsus, a, um, leafy (*frons, dis,* leaf).

91

Gazānia, called after Theodore of Gaza, fifteenth-century Greek scholar.

glabriusculus, a, um, rather glabrous (diminutive of *glaber*).

globulifer, a, um, *globulus,* little ball, and *fero,* bear, referring to the sporocarps.

Groenlandia, called after Johannes Groenland of Paris.

Hēbē, goddess of youth.

hederāceus, a, um, ivy-leaved.

Hēmerocallis, add: referring to the short-lived, beautiful, flowers.

hirsūtus, a, um, with long, distinct, not matted, hairs.

hirtellus, a, um, with short hairs (diminutive of *hirtūs*).

hispidus, a, um, bristly.

homoio, add: *Ranunculus homoiophyllus* has all leaves similar—no dissected ones.

hybernus, see *hībernus*.

hypnoïdēs, moss-like, resembling HYPNUM.

Hypochoeris, name, in Theophrastus, of *Hypochoeris radicata.* Sprague suggested ὑπώ, beneath, and χοῦρος, pig, referring to the bristles on the back of the phyllaries in some forms of the plant.

inaequidens, badly formed word meaning having unequal teeth.

infirmus, a, um, weak, feeble.

insectifer, a, um, insect-bearing (scientific Latin).

intactus, a, um, add: the flowers of *Neotinea intacta* do not open widely and are often self-pollinated. The plant may be considered a *virgo intacta.*

interjectus, a, um, interposed. *Polypodium australe* is diploid; *P. vulgare* tetraploid; and *P. interjectum* hexaploid and shows morphological characters intermediate between the other two.

Ipheion, origin obscure; ἰφυόν was the name in Theophrastus of a plant variously identified.

junceiformis, e, resembling *Agropyrum junceum.*

Juncus, probably connected with *jungo,* bind, from use in wickerwork, etc.

Lāser, the Latin name of *Ferula foetida* and allied plants.

latericīus, a, um, brick red, lit. made of bricks (*later, eris,* brick).

lautus, a, um, washed (elegant) (p.part. of *lavo,* wash).

lepidocarpus, a, um (*Carex tumidicarpa* Anderss. = *C. demissa* Hornem.).

limbospermus, a, um, *limbus,* border, and σπέρμα, seed. *Thelypteris l.* has the sori close to the margin of the segment.

Locusta, in botanical Latin means spikelet. Applied to *Valerianella l.,* from the form of the partial inflorescences.

Lolium, probably *L. temulenlum.* (Cf. Italian *loglio.*)

lūtescens, becoming yellow (*luteus,* yellow).

Lysichiton, λύσις, freeing, loosing, and χῐτών, tunic, referring to the lax spathe.

madritensis, e, of Madrid.

marginātus, a, um, having a conspicuous margin, as the achene of *Ranunculus marginatus* and seed of *Spergularia marginata* (*S. media*).

mãs, maris, male.

Matteucia, named, by Todaro, in honour of Carlo Matteuci, celebrated physicist.

megalo-, large (μεγάς, μεγάλη, μέγα).

megalūra, megalo- and οὐρά, tail, referring to the slender panicle of *Vulpia megalura.*

micranthus, a, um, having small flowers.

mikaniōïdes, resembling *Mikania scandens,* herbaceous composite climber of the New World, widely naturalized.

Mōrāceae, from genus MORUS (*morus,* mulberry tree).

mūtābilis, e, changeable (*muto,* change).

Najas, see *Nāias.*

Nardūrus, from NARDUS and οὐρά, tail, the narrow, tail-like, inflorescence resembles that of *Nardus stricta.*

nerteroiōdēs, resembling plants of the genus NERTERA (*Rubiaceae*).

Nicandra, called after Nicander, Greek author who wrote on antidotes.

nīliacus, a, um, of the river Nile.

nōdiflōrus, a, um, with flowers produced apparently at the nodes.

norvegicus, a, um. Norwegian.

ob-, prefix, usually meaning 'inversely' in botanical Latin.

octo-, eight-.

officīnārum, of the druggists' shops (gen.plur. of *officina*).

Oleāria, called after Adam Olearius (Oelenschläger), German botanist and librarian, born 1600.

Onoclēa, from ὄνος, beaker, wine-cup, and κλείω, shut close. The groups of sori are covered by an incurved leaf-margin forming bead-like structures (ὄνοκλια in Dioscorides was a synonym of ἄγχουσα). See ANCHUSA.

93

Ööcarpus, a, um, with egg-shaped fruit (ὠόν, egg, and καρπός, fruit).

orbiculāris, e, with circular outline.

Orthilia, origin obscure, the first element is perhaps from ὀρθός, straight, referring to the style.

pallidus, a, um, pale.

patientia, add: *lapazio* is the Italian name of *Rumex patienta.*

Pelisseriānus, a, um, of Guillaume Pellicier, sixteenth-century bishop of Montpelier, said, by Tournefort, to have discovered *Linaria P.* and *Teucrium scordium.*

pēs-caprae, *pēs*, foot, and *caprae*, gen.sing. of *capra*, nanny-goat, from the form of the leaflets of *Oxalis pes-caprae.*

Peucedanum, name of *Peucedanum officinale*, perhaps from πεύκη, PINUS sp. referring to resinous juice.

Phlomis, name, probably of PHLOMIS and VERBASCUM spp. with tomentum suitable for wicks (perhaps connected with φλόξ, flame).

Phormium, φορμίον, diminutive of φορμός, basket, from use of leaf-fibre.

Phragmītēs, add: the allied *Arundo donax* is still used for hedges in the Mediterranean Region.

Phuopsis, valerian-like, from φοῦ, VALERIANA, and ὄψις, appearance.

phyllanthes, φύλλον, leaf, and ἄνθησις, flowering, referring to the predominantly green perianth. Name perhaps suggested from the plant growing in Phyllis Wood, its *locus classicus*. The genus PHYLLANTHUS (*Euphorbiaceae*) is so called because the flowers of some species are borne on the ᵉdge of phyllodes.

Phȳsalis, from φυσαλλίς, bladder, bubble, referring to the inflated calyx.

Phȳsocarpus, φῦσα, bellows, and καρπός, fruit, referring to the inflated follicles.

Phytolacca, φυτόν, plant, and 'lac', a resinous substance secreted by the lac insect. The fruits yield a dye.

Pittosporum, from πίττα (Attic) pitch, and σπόρος, seed, from the resinous fluid surrounding the seeds.

Platanus, πλάτανος, the name of *P. orientalis* (πλατύς, broad, from the crown of the tree).

platyglossus, a, um, πλατύς, broad, and γλῶσσα, tongue, referring to the wide labellum.

Pōcilla, diminutive of *pōculum*, cup, referring to the seeds, cf. *Omphalospora.*

Pontedēria, called after G. Pontedera, Italian botanist.

Prīonītis, πριονῖτις, a name of *Stachys Alopecurus*, applied to FALCARIA because of its sharply serrate leaves (πρίων, ονος, saw, and *-itis*, feminine of *-ītēs*, q.v.)

purpurellus, a, um, diminutive of *purpureus*, referring to small plant.
pȳramidātus, a, um, pyramidal (πῡραμίς, pyramid).

Ramischia, called after Professor Fr. X. Ramisch of Prague (1798–1859), who worked on parthenogenesis in *Mercurialis annua*.
ramosissimus, a, um, superlative of *ramosus*.
Raphanistrum, see *-aster*.
rāriflōrus, a, um, with scattered flowers.
rĕclinātus, a, um, add: as the deflexed pod of *Ononis reclinata*.
rectus, a, um, straight (p.p. of *rego*, rule).
recutītus, a, um, circumcised, referring to the conical receptacle and reflexed rays of *Matricaria recutita* (*re-*, back, and *cutis*, skin).
Rhodotypos, ῥόδον, rose, and τύπος, pattern, model, referring to the flowers.
rigens, stiff, rigid (pres.p. of *rigeo*).
rōbur, oak, oak-timber.
rostrātus, a, um, beaked (*rostrum*, beak).
rotundus, a, um, round, circular, spherical.
rūgōsus, a, m, wrinkled.

sabrīnae, of the river Severn (*Sabrīna*).
salicētōrum, of willow thickets, see SALIX and *-ētōrum*.
Salpichroa, σάλπιγξ, ιγγος, trumpet, and χρῶς, skin, complexion, referring to the form and texture of the corolla.
samius, a, um, of the island of Samos.
sarachoīdēs, resembling the genus SARACHA, which was called after the Spanish botanist Isidor Saracha.
Scutellātus, a, um, having a scutella, see SCUTELLARIA, referring to the flat fruit of *Veronica scutellata*.
Seloānus, a, um, called after Friedrich Sellow, German botanist and traveller, who discovered *Cortaderia selloana*.
sparsiflorus, a, um, with scattered leaves (botanical Latin).
Spartium, σπαρτίον, name of several plants used for binding, including *Spartium junceum* (diminutive of σπάρτον, rope, cable).
spēluncārum, of caves (genit.plur. of *spelunca*, cave).
spīcant, origin obscure. C. Bauhin suggests derivation from *indica spica* (*Nardostachys jatamansi*, spikenard) because of similar root system. See NARDUS.
spīrālis, e, spiral (medieval Latin).
Staphylēa, *staphylodendron*, name, in Pliny, of *S. pinnata*, from σταφυλή, bunch of grapes (raceme), and δένδρον, tree, referring to the inflorescence.

95

Sternbergia, called after Count de Sternberg (1761–1838), botanist of Czechoslovakia.

stolōnifer, a, um, bearing stolons (*stolo, ōnis,* shoot, sucker, and *fero,* bear).

Struthopteris, στρουθός (sparrow), ostrich, and πτέρις, fern. The fertile fronds resemble ostrich feathers.

sūbulātus, a, um, awl-shaped (*sūbula,* awl, Rich, p. 629).

Tagetes, called after Tages, Etruscan divinity, grandson of Jupiter.

tangūticus, a, m, of Tangut, a district of the province of Kansa, China.

Tellima, anagram of MITELLA, an allied genus so called from its turban-like fruit (Rich, p. 426).

tener, era, erum, delicate, tender.

Thēlycranīa, θηλυκράνεια, name in Theophrastus of *T. sanguinea,* from θῆλυς, female, and κράνεια, *Cornus mas.*

Tolmiea, called after Dr W. F. Tolmie of Hudson Bay, who died in 1886.

triangularivalvis, e, the element *valvis* refers to the fruiting perianth segments of *Rumex t.*

tricornūtus, a, um, with 3 horns (*tri-,* three, and *cornutus,* horned).

Trillium, τρι-, 3, and LILIUM—lily-like plant with parts in threes.

Tripleurospermum, τρι-, three, πλευρόν, rib, and σπέρμα, seed, referring to the 3 ribs on the achene.

Trȳgonion, τριγόνιον, diminutive of τρυγών, turtle dove, *Streptopelia turtor* referring to *Geranium columbinum.*

Turrīta, add: **turrītis.**

ūnilaterālis, e, one-sided, of leaves, leaning to one side of the stem (*ūnus,* one, and *latus, eris,* side).

uniolōīdes, resembling the American genus UNIOLA.

varius, a, um, variable, changing.

villōsus, a, um, with long, rough hairs, shaggy.

viridis, e, green.

vīticella diminutive of *vītis,* the Latin name of the grape vine, *Vitis vinifera.*